Methods in Approximation

Mathematics and Its Applications

Methods in Approximation

Techniques for Mathematical Modelling

Richard E. Bellman

*Department of Electrical Engineering, University of Southern California,
Los Angeles, U.S.A.;
Center for Applied Mathematics, The University of Georgia,
Athens, Georgia, U.S.A.*

and

Robert S. Roth

Boston, U.S.A.

D. Reidel Publishing Company

A MEMBER OF THE KLUWER ACADEMIC PUBLISHERS GROUP

Dordrecht / Boston / Lancaster / Tokyo

Library of Congress Cataloging in Publication Data

Bellman, Richard Ernest, 1920–
    Methods in approximation.

    (Mathematics and its applications)
    Includes bibliographies and index.
    1.   Approximation theory.   I.   Roth, Robert, 1930–
II.   Title.   III.   Series: Mathematics and its applications
(D. Reidel Publishing Company)
QA221.B36   1986   511'.4   86–461
ISBN 90-277-2188-2

Published by D. Reidel Publishing Company,
P.O. Box 17, 3300 AA Dordrecht, Holland

Sold and distributed in the U.S.A. and Canada
by Kluwer Academic Publishers,
190 Old Derby Street, Hingham, MA 02043, U.S.A.

In all other countries, sold and distributed
by Kluwer Academic Publishers Group,
P.O. Box 322, 3300 AH Dordrecht, Holland

2–0388–200 ts

CONTENTS

EDITOR'S PREFACE

Approach your problems from the right end
and begin with the answers. Then one day,
perhaps you will find the final question.

'The Hermit Clad in Crane Feathers' in R.
van Gulik's *The Chinese Maze Murders*.

It isn't that they can't see the solution. It is
that they can't see the problem.

G.K. Chesterton. *The Scandal of Father
Brown* 'The point of a Pin'.

Growing specialization and diversification have brought a host of monographs and
textbooks on increasingly specialized topics. However, the "tree" of knowledge of
mathematics and related fields does not grow only by putting forth new branches. It
also happens, quite often in fact, that branches which were thought to be completely
disparate are suddenly seen to be related.

Further, the kind and level of sophistication of mathematics applied in various
sciences has changed drastically in recent years: measure theory is used (non-
trivially) in regional and theoretical economics; algebraic geometry interacts with
physics; the Minkowsky lemma, coding theory and the structure of water meet one
another in packing and covering theory; quantum fields, crystal defects and
mathematical programming profit from homotopy theory; Lie algebras are relevant
to filtering; and prediction and electrical engineering can use Stein spaces. And in
addition to this there are such new emerging subdisciplines as "experimental
mathematics", "CFD", "completely integrable systems", "chaos, synergetics and
large-scale order", which are almost impossible to fit into the existing classification
schemes. They draw upon widely different sections of mathematics. This programme, Mathematics and Its Applications, is devoted to new emerging
(sub)disciplines and to such (new) interrelations as exempla gratia:

- a central concept which plays an important role in several different mathematical
  and/or scientific specialized areas;
- new applications of the results and ideas from one area of scientific endeavour
  into another;
- influences which the results, problems and concepts of one field of enquiry have
  and have had on the development of another.

The Mathematics and Its Applications programme tries to make available a careful
selection of books which fit the philosophy outlined above. With such books, which
are stimulating rather than definitive, intriguing rather than encyclopaedic, we hope
to contribute something towards better communication among the practitioners in
diversified fields.

Because of the wealth of scholarly research being undertaken in the Soviet
Union, Eastern Europe, and Japan, it was decided to devote special attention to the
work emanating from these particular regions. Thus it was decided to start three
regional series under the umbrella of the main MIA programme.

As the authors stress in their preface there are two kinds of approximations involved when doing mathematical investigations of real world phenomena. First there is the idealized mathematical model in its full complexity and then there are the approximations one must make in order to be able to do something with it. And, of course, the inaccuracies adhering to the determination of whatever measured constants are involved. Thus,

> "In every mathematical investigation the question will arise whether we can apply our results to the real world .... Consequently, the question arises of choosing those properties which are not very sensitive to small changes in the model and thus maybe viewed as properties of the real process.
>
> V.I. Arnol'd, 1978

Such are the considerations which lie at the basis of several fields in mathematics: deformation theory, perturbation theory and approximation theory. In the setting above the task of approximation theory becomes that of finding that approximate model which precisely captures those properties which are not very sensitive to small changes. This also helps to make it clear that much more is involved than neglecting small terms or doing an expansion in some ϵ or throwing away all non-linear terms. Often indeed there will not even be an obvious small quantity.

The senior author of this volume, the late Richard Bellmann, has spent much time on thinking about the why and how of approximation and has pioneered several new methods in the field. The book can be accurately described as a survey of his thoughts on the topic during the last 25 years. Much of the material dates from the middle and late seventies. Here this material is presented in a coherent fashion by Bellmann and Roth without losing the typical inventive and stimulating character of much of Bellmann's writing.

The unreasonable effectiveness of mathematics in science ...

      Eugene Wigner

Well, if you know of a better 'ole, go to it.

      Bruce Bairnsfather

What is now proved was once only imagined.

      William Blake

As long as algebra and geometry proceeded along separate paths, their advance was slow and their applications limited.

But when these sciences joined company they drew from each other fresh vitality and thenceforward marched on at a rapid pace towards perfection.

Joseph Louis Lagrange.

Bussum, December 1985   Michiel Hazewinkel

PREFACE

Any attempt by the applied mathematician to model a physical observation must involve approximation. Since the motivation for such modelling is to be able to predict accurately the behavior of a system under study, it becomes very important to understand the underlying approximation assumptions for any given problem. It is well understood that each problem has its own set of unique assumptions.

From a classical point of view, modelling of physical phenomenon has involved a very strict discipline of combining experimental observation with careful mathematical construction to obtain a consistent blend of experimentation and calculation which, in the end, predicts result consistent with observation. This ,very often, is the goal of applied mathematical research.

It is our contention that the search for accurate mathematical modelling is fundamentally intertwined with the ideas of approximation. We wish to clearly distinguish between two types of approximation: the physical approximation and the mathematical approximation. The modelling of any physical phenomenon involves creating a mathematical structure in which physical subunits are modelled in terms of mathematical expressions. Such expressions are approximations in the sense of characterizing certain interesting behavior. For example, in elasticity, the constitutive relations are often given as linear expressions while a more accurate plasticity model demands that nonlinear terms be added. The consequence of this

type of approximation is the derivation of
often a very complex set of final equations
governing the behavior of the system. Since
these equations may be difficult to solve ,
mathematical approximation may be required to
obtain a solution.

The present volume is an accumulation of our
thoughts on methods of modern mathematical
techniques of approximations over the past several years. It is not our intention to survey
the vast amount of work in this area, but rather to concentrate on selective topics in this
challenging and rapidly expanding area of applied mathematical study.

Since the construction of mathematical approximation must have an underlying foundation,
we devote chapter 1 to the ideas which are central to its construction. We consider the abstract vector space for it allows us to define
approximation errors with precision.
This,then,is our starting point.

Chapter 2 discusses polynomial approximation
and we are particularly interested in studying
curve fitting by segmented straight lines.
These simple ideas require dynamic programming
techniques for a solution. At the other end of
the spectrum, we consider a three dimensional
approximation which is used in the popular finite element method.

The more general ideas of polynomial splines
are treated in chapter 3. Here polynomial approximation offers a very smooth representation
of a function $f(x)$ and therefore allows us to
store only the coefficients of the spline if we
wish to reproduce $f(x)$, thereby saving enormous
space in computer applications.

Chapter 4 discusses the ideas of quasilinearization allowing us not only to solve nonlinear differential equations in an efficient
way, but also to determine approximate parame-

ters and initial conditions of the differential equations if, somehow, the behavior of the system can be observed and measured.

Differential approximation, the subject of chapter 5, is a slightly different approximation. Here we are given a nonlinear differential equation and we seek to determine an approximate linear differential equation whose exact solution is a good approximation to the original solution in the range of interest.

Differential quadrature is considered in chapter 6. Here we return to numerical techniques to allow us to find approximate solutions to partial differential equations of the form,

$$u_t(x,t) = g(x,t,u(x,t),u_x(x,t))$$

$$u(x,0) = h(x),$$

The same approximation techniques can be used to solve higher order systems and using blends of this with quasilinearization, we can consider solving systems with partial information.

Chapter 7 discusses exponential approximation where we seek approximations of the form,

$$u(t) = \sum_{n=1}^{N} a_n e^{\lambda_n t}.$$

These techniques are used to convert the Fredholm integral equation,

$$u(t) = f(t) + \int_0^1 k(|t - s|)u(s)\, ds,$$

into a set of differential equations representing an interesting two point boundary value problem.

Chapter 8 applies the approximation techniques to a study of the Ricatti equation which continually arises in mathematical physics.

$$u' + u^2 + p(t) u + g(t) = 0 \quad ,$$

(1)

$$u(0) = c.$$

Using the ideas of the maximum operation, we can characterize the solution of the Ricatti equation by means of upper and lower bounds. We can approach the solution of Eq.(1) yet another way ,via quasilinearization and take full advantage of quadratic convergence in the numerical solution.

The solution of approximate equations is discussed in chapter 9 . The idea here is to replace a difficult equation whose solution is unknown with an approximate equation whose solution is known exactly. These concepts are clearly based on classical techniques in applied mathematics as we point out by first considering perturbation methods in solving nonlinear differential equations. Here we have a precise way of defining a sequence of linear differential equations whose solutions are well known. Thru the perturbation series, the nonlinear equation is systematically approximated by a simpler set of linear equations.

Modern approximation methods such as the polynomial spline affords us a new freedom in selecting our problems and the theory of dynamic programming,for example, allows us to bring together many disciplines in defining the approximate equations themselves.

The goal of this chapter is to aquaint the reader with several of these techniques which may have immediate use in solving complex problems in applied mathematics.

Finally, chapter 10 discusses an application of the finite element method to compute the three dimensional magnetic field determination. In this example we use the polynomial approximation developed in chapter 2 to solve the problem.

In view of the computer revolution and in particular the interest in the small personal computer, solving difficult problems by approximation techniques has wide appeal. This is evident in the necessity of simplifying the mathematics thru approximation techniques. An interesting by product of the ideas on approximation is the need to develop new ways to store information. While this is discussed only briefly in the text, situations will continually arise where large storage, beyond the capacity of the computer, is required for solution.In these cases, approximation techniques can serve us very well.

The application of approximation techniques is itself a very challenging endeavor. With careful thought, the mathematician can undoubtedly develop ideas far beyond the scope of this book. We sincerely hope this is true and we look confidently into the future.

Richard E. Bellman

Robert S. Roth

Santa Monica, California
Cambridge,Massachusetts

Chapter 1

BASIC CONCEPTS

1.1   INTRODUCTION

A successful approximation has the following three properties:

(a)   A well defined mathematical problem

(b)   A rational method for deriving and approximating problems and/or solutions ,

(c)   A measure of the error incurred in using these approximations .

In this chapter we shall consider all three properties to varying degrees in our study of new methods of approximation. At the outset we wish to point out that methods of approximation run the range from classical techniques to modern ideas and we intend to intersperse both extremes throughout the text.

In general the mathematical problems we intend to study will derive from physical and well defined mathematical situations, therefore property (a) can be studied by classical methods. Our primary interest will be to consider properties (b) and (c) of a successful approximation. We shall begin by introducing the abstract concepts which will form the structure on which our approximation ideas are based. The properties of the Integral Domain, the Field

and the Vector Space serves us well for they
establish a rigorous framework within which
much can be done.

1.2   INTEGRAL DOMAINS,FIELDS AND VECTOR SPACES

All ideas of approximation should lie within a unified algebraic structure. The structure
should be general enough to comply with properties (a), (b) and (c) of a successful approximation.   We begin with the Integral Domain
which is constructed in the following way.

Definition:

Let A be a set of elements a, b, c ...
for which a + b and the product ab of
any two elements a and b (distinct or not)
of A are defined. Then A is called an
Integral Domain if the following postulates (1) - (9) hold:

(1) Closure: If a and b are in A,then the
sum a + b and the product ab are also
in A,

(2) Uniqueness: If a = a', and b = b'
in A,then a+b = a'+b', and ab = a'b',

(3) Commutativity: for all a and b in A,
a+b = b+a, ab = ba,

(4) Associativity: for all a,b and c in
A, a+(b+c) = (a+b)+c, a(bc) = (ab)c,

(5) Distributivity: for all a,b and c in
A, a(b + c) = ab + ac,

(6) Zero: A contains an element 0, such
that a + 0 = a for all a in A,

(7) Unity: A contains an element 1 not
zero such that a 1 = a for all a in A,

(8) Additive Inverse: for each element a
    in A ,the equation a + x = 0 has a
    solution in A,

(9) Cancellation: if c is not 0 and
    ca = cb, then a = b.

With this definition of an Integral Domain,a
Field can be defined.

Definition:

A field F is an Integral Domain
which contains for each a ≠ 0, an inverse
element a^{-1} satisfying the relation $a \, a^{-1} = 1$.

Within the context of a vector space, fields
play the role of a structured set of scalars.

Definition:

A vector space V over a field F is a set of
elements called vectors such that any two vectors α and β of V determine a (unique) vector α
+ β as sum and that any vector α from V and c
from F determine a vector cα in V with the
properties:

(1) Associativity:
    $\alpha + (\beta + \psi) = (\alpha + \beta) + \psi$

(2) Identity: There is a vector 0 in V
    such that, α + 0 = α for all α,

(3) Inverse Law: There is a solution in
    V to the equation α + x = 0, that
    is in V,

(4) Commutative Law: for all α and β
    in V, α + β = β + α,

(5) Distributive law:
    c(α + β) = cα + cβ

(6) (ab)α = a(bα);

(7) Unity:
$$1\alpha = \alpha$$

We now have a vector space in which to work. Our particular interest is to consider all functions $f(x)$ whose domain is S and whose range is a field F assigning to each x in S a value $f(x)$ in F. It is easy to show that the set of all such functions f forms a vector space over F if the sum $h = f + g$ and the scalar product $h' = cf$ are functions defined for each x in S by the equations $h(x) = f(x) + g(x)$ and $h'(x) = c\, f(x)$.

Hence all the rich results derived from the definition of the vector space are available to us in our development of approximation techniques. Furthermore, by structuring our development on the vector space, applications beyond the scope of this book may be apparent to the reader.

1.3 SUBSPACES, BASES AND INNER PRODUCTS

A subspace S of a vector space V is a subset of V which is itself a vector space with respect to the operations of addition and scalar multiplication.

For example, for a given set of vectors $a_1, a_2, \ldots a_n$ the set of all combinations,

$$c_1 \alpha_1 + c_2 \alpha_2 + \ldots + c_m \alpha_m$$

is a subspace of V. This is because of the identities:

$$(a_1 \alpha_1 + \ldots + a_m \alpha_m) + (b_1 \beta_1 + \ldots + b_m \beta_m) =$$

$$= (a_1 + b_1)\alpha_1 + (a_2 + b_2)\beta_2 + \ldots + (a_m + b_m)\beta_m,$$

$$a_1(b_1\alpha_1 + \ldots + b_m\alpha_m) = (a_1 b_1)\beta_1 + \ldots + (a_m \beta_m)\alpha_m$$

showing that the distributive law holds for all vectors in the subspace

We can also prove that the intersection A , B of any two subspaces of a vector space V is itself a subspace of V.

We now make the observation that if V is a vector space and A is a subspace of V and a vector α is chosen in V, then if we wish to approximate α by a choosing the 'closest' vector β in A, then the "error" in the approximation must lie in the subspace V - A. If the set V - A is 0, (the empty set), then the approximation is exact (α = β). By defining our concepts of "closest" and "error" carefully, we can further refine the concept of approximation. Let us consider the idea of linear independence.

The vectors $\alpha_1, \alpha_2, \ldots \alpha_3$ are linearly independent over the field F if and only if for all scalars c_i in F,

$$c_1\alpha_1 + c_2\alpha_2 + \ldots + c_m\alpha_m = 0 \qquad (1-1)$$

implies that $c_1 = c_2 = \ldots = c_m = 0$ and the vector space has the dimension m.

From the idea of linear independence we introduce the bases of a vector space as a linearly independent subset which spans the entire space. We wish to impose on this vector space a means of defining lengths of vectors which play a central role in considering modern methods of approximation.

In an n-dimensional space, the inner product

of two vectors $\alpha(a_1, a_2 \ldots a_n)$ and $\beta(b_1, b_2, \ldots b_n)$ with real coefficients a_1 etc. defines the quantity,

$$(\alpha, \beta) = a_1 b_1 + a_2 b_2 + \ldots + a_n b_n, \qquad (1\text{-}2)$$

sometimes written as $\alpha \bullet \beta$.

   As a direct consequence of the definition, we see immediately,

$$(1) \quad (\alpha + \beta, \gamma) = (\alpha, \gamma) + (\beta, \gamma),$$

$$(2) \quad (\alpha, \beta) = (\beta, \alpha) , \qquad\qquad (1\text{-}3)$$

$$(3) \quad (\alpha, \alpha) > 0 , \quad \text{if } \alpha \neq 0.$$

Such a positive function defined over a vector space allows us to define the length of a vector α as

$$|\alpha| = ((\alpha, \alpha))^{1/2} .$$

   We choose to take the square root of (α, α) to preserve the dimension of length. In fact any function of vectors α, β, satisfying Eqs. (1-1, 1-3). may serve to define distance. An observation which we shall make valuable use of later.

1.4   SPACES, SUBSPACES AND APPROXIMATION

Let V_n be an n-dimensional vector space with
bases $(\alpha_1, \ldots \alpha_n)$.   Since the bases spans
the space any vector β, which is a member
of V_n, can be written as a linear combination
of the base vectors. So,

$$\beta = a_1 \alpha_1 + a_2 \alpha_2 + \ldots + a_n \alpha_n .$$

We now ask a trivial but important question.
Is there another vector γ in V_n which best
approximates β?
Let

$$\gamma = c_1 \alpha_1 + c_2 \alpha_2 + \ldots + c_n \alpha_n,$$

which we can surely do knowing the base vec-
tors.We can then define an error vector,

$$(\beta - \gamma) = (a_1 - c_1) \alpha_1 + (a_2 - c_2) \alpha_2 + \ldots$$

$$+(a_n - c_n)\alpha_n .$$

If we ask for the best approximation to
the vector β, clearly our choice would be,
$a_1 = c_1, a_2 = c_2 \ldots a_n = c_n$ but what happens
if we restrict γ to be in a subspace V_m of V_n,
$m < n$? The answer to the previous question is
no longer trivial.

Again we can define an error vector,

$$(\beta - \gamma) = (a_1 - c_1)\alpha_1 + \ldots + (a_m - c_m)\alpha_m$$

$$+a_{(m+1}\ \alpha_{(m+1)}\ +\ \dots\ +a_n \alpha_n \ .$$

The standard technique is to select $(c_1 \dots c_n)$ such that,

$$|\alpha - \gamma| = (\ (\alpha-\gamma,\alpha-\gamma))^{1/2} ,$$

is a minimum. Of course, the minimizing vector γ is the projection of β into the subspace V_m, yet in many cases the computation may not be easy.

By viewing approximation through the structure of the vector space, we hope to make clear the intricacies of modern approximation techniques.

1.5   THE CONTINUOUS FUNCTION

Throughout this book we shall be dealing with approximation techniques as they apply to continuous functions $f(x)$ defined over the range (a,b). Since any finite range can be transformed into the range $(0,1)$ by a simple transformation, we shall consider only the range $(0,1)$ for the remainder of this chapter. We note, without proof, the following result. The set of all continuous functions $f(x)$ defined over the closed interval, $0 \leq x \leq 1$, form a vector space under vector addition and scalar multiplication, ie. they satisfy all the postulates (1-9) of a vector space given in (1.2)

In this vector space we can define the inner product, or $L(p)$ norm, $p>=1$, in the following way. Let $f(x),q(x)$ be continuous functions over $(0,1)$. We let,

$$(f,g) = \left(\int_0^1 |f(x) - g(x)|^p \, dx \right)^{1/p},$$

for convenience, we choose the L(2) norm, where

$$(f,g) = \left(\int_0^1 |f(x) - g(x)|^2 \, dx \right)^{1/2}.$$

Over this vector space, we are at liberty to define other norms when it is convenient to do so.

1.6   POLYNOMIAL SUBSPACES

Several subspaces of V_f need to be mentioned for it is in these that we intend to do much of our work. Let V_N be the set of all polynomials $p_n(x)$ $n < N$ defined over (0,1). The space V_n can be shown to be a subspace of V_N. Moreover, the same norm is valid in V_N. Thus if $p_n(x)$ and $q_m(x)$ are members of V_N, m<N, then,

$$(p_n, q_m) = \left(\int_0^1 |p_n(x) - q_m(x)|^2 \, dx \right)^{1/2}.$$

The entire space V_n can be generated by N linearly independent polynomials, that is, any polynomial $r_n(x)$ $n \leq N$ can be expressed as a linear combinations of the base polynomials.

1.7   SPACES GENERATED BY DIFFERENTIAL
EQUATIONS

Consider the linear partial differential equation of the second order,

$$a_1 \frac{\partial^2 f}{\partial x^2} + a_2 \frac{\partial^2 f}{\partial y^2} + a_3 \frac{\partial^2 f}{\partial x \partial y} + a_4 \frac{\partial f}{\partial x}$$

$$(1-4)$$

$$+ a_5 \frac{\partial f}{\partial y} + a_6 f = 0 ,$$

where a_i (i=1,...,6) are continuous functions of the variables x, y in a closed region G bounded by a simple curve C.

For boundary conditions one may require the value of the function f(x,y) on the closed curve (Dirichlet Condition) or the value of the normal on the curve (Neumann Condition).

Suppose t is a parameter which is defined on the curve C. Eq.(1-4) admits for every $\zeta(t)$ which is continuous with continuous deriva- tives, a solution of Eq.(1-4) which reduces to $\zeta(t)$ on C. Let F be the set of all solutions of Eq.(1-4). We wish to show F is a vector space. If we follow the postulates of a vector space, we see,

Closure: Let f(x,y) and g(x,y) satisfy Eq.(1-4), then because it is linear, f(x,y)+g(x,y) is also a solution of Eq.(1-4) and therefore is in F.

(1)   Associativity: Clearly   for f(x,y),g(x,y) and h(x,y) in F, we have f(x,y) + (g(x,y) + h(x,y)) = (f(x,y) + g(x,y)) + h(x,y).

(2)   Identity: Because   Eq.(1-4)   is
homogeneous, $f(x,y)= 0$ everywhere is also a solution of Eq.(1-4) and is the identity element
under addition.

(3) Inverse Law: For each $f(x,y)$ in F,
$f(x,y)$ is also in F because Eq.(1-4) is homogeneous and therefore every element in F has an
inverse in F.

(4) Commutivity: Clearly, $f(x,y) + g(x,y) =$
$g(x,y) + f(x,y)$.   We can further say that F is
an abelian group under addition

We introduce scalar multiplication by noting
that if $f(x,y)$ is in F, c $f(x,y)$ is also in F
for all c real (or complex). We quickly note
that this is true because Eq.(1-4)   is,as we
know, homogeneous.

(5) Distributivity;

$$(a + b) \; f(x,y) = a \; f(x,y) + b \; f(x,y)$$

$$a(f(x,y) + g(x,y)) = a \; f(x,y) + a \; g(x,y)$$

(6) and, $(ab)f(x,y) = a(b \; f(x,y))$

(7) Unity:

$$1 \; f(x,y) = f(x,y)$$

Therefore F is a   vector space of continuous
functions defined over A.   Furthermore we can
introduce a norm,

$$(f,g) = \left(\int_A |f(x,y) - g(x,y)|^2 \; dA \right)^{1/2}. \qquad (1-5)$$

to form a normed vector space. Hence the element of F are subject to all the theorems for
normed vector spaces, many of which we will use
explicitly in our approximations.   We wish to

point out that again F is a subspace of the set
of all continuous functions defined on A. The
idea immediately comes to mind of approximating
any function g(x,y) by the "best" function
f(x,y) in F. By that we mean choosing f(x,y)
such that,

$$(f,g) = \min_{f \text{ in } F} \ \int_A |f(x,y) - g(x,y)|^2 \, dA)^{1/2}.$$

We may ask a more fundamental question.
What choices can we make for the coefficients
$(a_1, a_2, .. a_6)$ in Eq.(1-4) to get the best
possible approximation to g(x,y) . We shall
be discussing the problem extensively in later
chapters.

A careful look at Eq.(1-5) reveals an interesting point. To get the best approximation to
g(x,y), we must be able to search the entire
space F. An immediate idea is to determine a
bases of F and search over all linear combinations of such bases.

To illustrate our point briefly. Consider
the second order ordinary differential equation.

$$\frac{d^2 f(x)}{dx^2} + a \frac{d f(x)}{dx} + b \, f(x) = 0 .$$

$$\text{(1-6)}$$

$$0 < x < 1$$

Now define a set of functions on (0,1)
which may be generated by two base functions
$f_1(x)$ and $f_2(x)$ which are solutions of
Eq.(1-6) with the following boundary conditions.

$f_1(x)$ where $f_1(0) = 1,$ $df_1(0)/dx = 0$

$f_2(x)$ where $f_2(0) = 0,$ $df_2(0)/dx = 1.$

In this case $f_1(x)$ and $f_2(x)$ are linear combinations of exponants which , themselves, are
good approximating functions. If, for example,
$a = b = 0$ in Eq.(1-6) , then $f_1(x) = 1$ and
$f_2(x) = x$ will generate the vector space
$g(x) = c_1 + c_2 x,$which we will consider in detail
in chapter 2 . Again, if $a = 1$ and $b = 1$ then
$f_1(x) = cosx$ while $f_2(x) = sinx$ and we have at
our disposal the vector space generated by
$g(x) = c_1 cosx + c_2 sinx$. The reader is urged to
study the Sturm-Liouville systems for further
results.

The linear differential equation,therefore,
can play a very important part in dealing with
approximation techniques. They offer a rich
area for development which we shall explore in
later chapters.

1.8   THE PIECEWISE LINEAR FUNCTION

In this section we shall consider the
piecewise linear function which we shall study
in detail in Chapter 2. Let $g(x)$ be a piecewise
linear function which is continuous in $0 \le x \le 1.$

Figure 1.1

Piecewise Linear Function

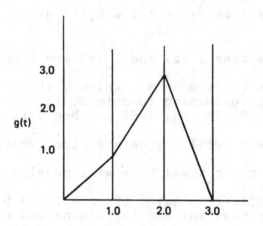

Let g(x) be given by defining the   values of
$g(x_i) = g_i$, where   both   the x_i and g_i are
known Again, this set of all piecewise linear
functions forms a vector space for the set g is
an abelian group under addition, ie. we have,

(1) closure: if f(x) g(x) are members of G,

so is f(x) + g(x),

(2) associativity: for all f(x),g(x) and h(x)
in G,

(f(x)+g(x)) + h(x) = f(x) +(g(x)+h(x))

(3) identity: g = 0 is a member of G,

(4) inverse: for every f(x) in G, -f(x)
is also in G,

(5) commutivity: for all f(x) and g(x) in G,

(f(x)+ g(x)) = (g(x)+f(x))

So G itself is an abelian group and under the
definition of scalar multiplication, becomes a
vector space , in fact, it is a subspace of the
space of all continuous functions. A norm can
be defined over G as the usual l(2) norm,

$$(f,g) = (\int_0^1 |f(x) - g(x)|^2 \, dx)^{1/2}.$$

We shall return to this later.

1.9   DISCUSSION

In the following chapters we intend to develop our ideas of approximation centered about
the structure of the abstract vector space. If
we are given a function f(x) continuous over
the interval (0,1), then our task will be to
explore the many subspaces in which an approximation to f(x) can be found to our advantage.

Any advantage in approximating a function
f(x) from an appropriately chosen subspace depends clearly on the application. We wish to
cite several examples. The approximation of
f(x) by a series of polynomials allows us to
study some of the gross properties of f(x) in
an analytic way. The computer storage of f(x)
over (0,1) is reduced to storing (n+1) coefficients of an n-th order polynomial approximation. We could ask that the approximating
polynomial be very simple,say linear, but we
wish to know how the intervals are optimally
subdivided into N subintervals to best approximate f(x).

We could take a very different approach by
seeking to approximate f(x) by a function g(x)
which itself is the solution of a known differential equation for which the coefficients must
be determined. This is a very neat way to con-
nect f(x) with a process, defined by the

differential equation, which will generate the
best approximation to f(x).

In the later chapters we shall consider the
approximation problem itself. Rather that trying to find an approximate solution to a difficult differential equation, we shall seek an
exact solution to an approximate equation. In
general we shall try to construct approximate
techniques within the context of the abstract
vector space.

1.10   BIBLIOGRAPY AND COMMENTS

A good introduction to the basic ideas of
modern algebra is found in,

Birkhoff,G. and S. MacLane :1953,
A Survey of Modern Algebra
Macmillan Company, New York

Gillman,L, and M.Jerison :1960,
Rings of Continuous Functions,
Springer

Chapter 2

POLYNOMIAL APPROXIMATION

2.1 INTRODUCTION

In this chapter we shall consider polynomial approximation in its most simple form. As in the last chapter we shall restrict ourselves, for convenience, to the closed interval (0,1), and we will let f(x) be a real valued continuous function defined in the interval.

Our first essential task is to show that the polynomial is a good candidate for approximating f(x) over (0,1). If this is true, the numerical ideas which follow will have a clear theoretical basis.

We can establish this by proving the Weierstrass Polynomial Approximation theorem. Let f(x) be a real valued continuous function defined in the closed interval (0,1).

We shall show that there exist a sequence of polynomials $P_n(x)$ which converge to f(x) uniformly as n goes to infinity. To begin we choose the polynomial $P_n(x)$ to be of the form,

$$P_n(x) = \sum_{m=0}^{n} C_{n\,m} \, f(m/n) \, x^m (1 - x)^{n-m} ,$$

where $C_{n\,m}$ is the combination of n things taken m at a time. Consider the binomial expansion ,

$$(x + y)^n = \sum_{m=0}^{n} C_{m\,n} x^m y^{(n-m)} \qquad (2-1)$$

by differentiating Eq.(2-1) by x and
multiplying the result by x, we have

$$n\,x(x + y)^{n-1} = \sum_{m=0}^{n} m\,C_{n\,m} x^m y^{(n-m)} \qquad (2-2)$$

and by differentiating Eq.(2-1) twice by x
and multiplying the results by x^2 we have,

$$n(n-1)x^2 (x + y)^{(n-2)} = \sum_{m=0}^{n} m(m-1)\,C_{n\,m} x^m y^{(n-m)}. \qquad (2-3)$$

Now if we set,

$$r_m(x) = C_{n\,m} x^m (1-x)^{n-m} , \qquad (2-4)$$

then, by using Eq,(2-1),Eq.(2-2) and Eq.(2-3),
setting y = 1-x, we have the following three
important properties,

$$\sum_{m=0}^{n} r_m(x) = 1$$

$$\sum_{m=0}^{n} m\,r_m(x) = n\,x$$

and

$$\sum_{m=0}^{n} m(m-1)r_m(x) = n(n-1)\,x^2 .$$

Using these properties, we can easily see that,

$$\sum_{m=0}^{n} (m-nx)^2 r_m(x) = \sum_{m=0}^{n} m^2 r_m(x) - 2n x \sum_{m=0}^{n} m r_m(x)$$

$$+ \sum_{m=0}^{n} n^2 x^2 r_m(x),$$

$$= (nx + n(n-1)x^2) - 2n^2 x^2 + n^2 x^2,$$

$$= nx(1 - x).$$

We now note that since $f(x)$ is continuous over the closed interval $(0,1)$, then there exists an M such that,

$$|f(x)| \leq M , \quad 0 \leq x \leq 1.$$

Also by uniform continuity, there exists, for every $\epsilon > 0$, $\delta > 0$, such that, if

$$|x - x'| < \delta, \text{ then } |f(x) - f(x')| < \epsilon.$$

Now , consider the expression,

$$|f(x) - \sum_{m=0}^{n} f(m/n) r_m(x)| = |\sum_{m=0}^{n}(f(x) - f(m/n)) r_m(x)|,$$

$$\leq | \sum_{|m - nx| < \delta n} (f(x) - f(m/n)) r_m(x)|,$$

$$+ | \sum_{|m - nx)| > \delta n} (f(x) - f(m/n)) r_m(x) | , \qquad (2\text{-}5)$$

but if we consider the first expression, we can see

$$| \sum_{|m - xn| < \delta n} (f(x) - f(m/n)) r_m(x) | \leq \epsilon \sum_{m=0}^{n} r_m(x) = \epsilon.$$

To get a bound on the second term in Eq.(2-5) , we observe that since $r_m(x) \geq 0$, over all m chosen in the second term,

$$|(m - nx)/\delta n| > 1 ,$$

so

$$((m - nx)/\delta n)^2 > 1.$$

Therefore, for all m , such that $|m - nx| > \delta n$,

$$\sum_{\substack{m \\ |m - nx| > \delta n}} r_m(x) \leq \sum_{\substack{m \\ |m - n x| > \delta n}} ((m - nx)/\delta n)^2 r_m(x).$$

Since $r_m(x) \geq 0$, we can combine the two expressions in Eq.(2-5) to give the useful result,

$$\sum_{|m - nx| \geq \delta n} r_m(x) \leq \sum_{m=0}^{n} ((m - nx)/\delta n) r_m^2(x).$$

Returning to the main proof, we can now observe that,

$$\left| \sum_{|m-nx|>\delta n} (f(x) - f(m/n)) r_m(x) \right|$$

$$\leq \left| \sum_{\substack{m \\ |m-nx|>\delta n}} f(x) r_m(x) \right| + \left| \sum_{\substack{m \\ |m-nx|>\delta n}} f(m/n) r_m(x) \right|,$$

$$\leq 2M \left| \sum_{\substack{m \\ |m-nx|>\delta n}} r_m(x) \right| ,$$

$$\le \quad 2M \sum_{m=0}^{n} ((m-nx)/\delta n)^2 r_m (x) ,$$

$$= \quad 2M/(n^2 \delta^2) \sum_{m=0}^{n} (m-nx)^2 r_m (x) ,$$

$$= \quad 2M/(n^2 \delta^2) \, nx(1-x) ,$$

$$\le \quad M/(2n\delta^2),$$

since $\max_{0<x<1} x(1-x) = 1/4.$

which approaches zero as n goes to infinity. Therefore we have the desired result, namely

$$|f(x) - \sum_{m=0}^{n} f(m/n)r_m (x) | < \epsilon,$$

and the polynomial $P_n (x) = \sum_{m=0}^{n} f(m/n)r_m (x)$ converges uniformly to $f(x)$ in the closed interval $(0,1)$.

In the next section we shall consider how we can best approximate $f(x)$ over $(0,1)$ by segmented straight lines, the simplest of polynomals.

2.2   PIECEWISE LINEAR FUNCTIONS

We shall now define a piecewise linear function in the following way. Over each interval g(x) is to be linear, therefore,

Let

$$g(x) = a_i + b_i x \; ,$$
$$u_0 = 0,$$
$$u_N = 1. \qquad u_{j-1} \le x \le u_j,$$
$$j = 1, \ldots, N-1.$$

Let G be the set of all piecewise linear functions in (0,1). The basic properties of G can immediately be written down,

- Closure: if $g_1(x)$ and $g_2(x)$ are piecewise linear functions then so is $g_1(x) + g_2(x)$.
- Associativity: $g_1(x) + (g_2(x) + g_3(x))$
  $= (g_1(x) + g_2(x)) + g_3(x)$
- Additive Identity: $g_0(x) = 0$ is a member of G,

- Additive Inverse: given $g_1(x)$ in G, then $-g_1(x)$ is also in G,

- Commutivity: $g_1(x) + g_2(x)$
  $= g_2(x) + g_1(x),$

Therefore G is an abelian group under addition. But we can go further.   Let a be a number in a field F. Therefore,

- $a\,g_1(x)$ is in G,

- $a(g_1(x) + g_2(x)) = a\,g_1(x) + a\,g_2(x),$

- $(a + b)\,g_1(x) = a\,g_1(x) + b\,g_1(x)$,

- $(ab)\,g_1(x) = a(bg_1(x)).$

So G is a vector space.

We now define a continuous sequence of piece-wise linear functions over the closed interval (0,1) $g_n(x)$ if it is continuous and has n joining points (called knots) within the interval (0,1). Furthermore $g_n(x)$ is a piecewise continuous approximation to f(x) if at each knot x_i, $g_n(x_i) = f(x_i)$.

We can now demonstrate the following result. If f(x) is uniformly continuous in the closed interval (0,1), then there exists a piecewise linear function $g_n(x)$ which converges uniformly to f(x) as n goes to infinity. Since f(x) is uniformly continuous for every $\epsilon > 0$, there exists a $\delta > 0$ such that,

$$\text{if } |x - x'| < \delta,$$

$$\text{then } |f(x) - f(x')| < \epsilon.$$

We can subdivide the interval (0,1) into N sub-divisions such that for all n > N, $|x_{n+1} - x_n| < \delta$. We can now construct a piecewise linear function $g_n(x)$ over (0,1). Therefore, within each sub-division,

$$|f(x) - g_n(x)| < |f(x) - ((\frac{f(x_{n+1}) - f(x_n)}{x_{n+1} - x_n})$$

$$(x - x_n))|,$$

$$< |f(x) - f(x_n))| + |\frac{f(x_{n+1}) - f(x_n)}{x_{n+1} - x_n}|$$

$$|x - x_n|,$$

$$< 2\epsilon.$$

Therefore, for any f(x) which is uniformly continuous on (0,1), there is a piecewise linear function $g_n(x)$ which is uniformly close to f(x) in the interval.

2.3   CURVE FITTING BY STRAIGHT LINES

We shall now use our results to solve an interesting problem in approximation. Consider the function f(x) which is defined on the closed interval (0,1). We wish to approximate this curve by a piecewise linear function $g_n(x)$ where n is not fixed and we determine the best position for the knots of $g_n(x)$.

Our motivation is to introduce a very simple approximation, the straight line, and to allow complexity by varying the position of the knots.

Figure 2.1

Piecewise Linear Approximation

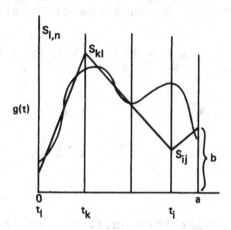

In classical problems of this type, the
approximation is found by least squares methods,
assuming that the locations of the knots are known.
We now wish to relax this restriction by allowing
the knot locations   to be unknown together with
the set $y_i = g(x_i)$.

We are now faced with a more complicated
problem whose solution can be defined in terms
of Dynamic Programming.   One method for determining the position of the knots to best approximate f(t) in the interval (0,1) may be
thought of as a multistage decision process. At
the Nth stage, after determining the best posi-
tions for N-1 knots,we seek to find how best to
place the Nth knot. This process is best il-
lustrated in the one dimensional case.

2.4   A ONE DIMENSIONAL PROCESS IN DYNAMIC PROGRAMMING

Let us consider the problem of determining the maximum of N functions of N variables. Let

$$R(x_1, x_2, \ldots x_N) = h_1(x_1) + h_2(x_2) + \ldots + h_N(x_N), \tag{2-6}$$

where

$$x_1 + x_2 + \ldots + x_N = x,$$

$$x_i \geq 0, \tag{2-7}$$

$$i = 1, \ldots N.$$

The particular form Eq.(2-6) is interesting,for it assigns to each x_i,a value $h_i(x_i)$ independent of the other values of x_j. Expression Eq.(2-7) puts a constraint on the choice of x_i in that only a limited amount of quanities,let us say resources, are available. With our approximation problem in mind, we can consider an interval $(0,x)$ on the real line as divided into N subdivisions where x_i is now the length of each subdivision. Then the constraints have the form,

$$x_1 + x_2 + \ldots + x_N = x.$$

2.5   THE FUNCTIONAL EQUATION

To treat the particular problem of maximizing (or minimizing) the function,

$$R(x_1, x_2, \ldots x_N) = h_1(x_1) + h_2(x_2) + \ldots + h_N(x_N),$$

subject to the constraint,

$$x_1 + x_2 + \ldots + x_N = x,$$

$$x_i \geq 0,$$

we imbed it in a family of problems. Instead of considering x and N to be fixed, we let x assume any positive value and N can assume any integer value.

Now we must say something about the functions $h_i(x_i)$. In our imbedded problem, we shall assume $h_i(x_i)$ is continuous and bounded over a large maximum interval (0,a), for i = 1,.,N. Therefore,

$$h(x) = h_1(x_1) + h_2(x_2) + \ldots + h_N(x_N),$$

$$x_1 + x_2 + \ldots + x_N = x < a$$

$$x_i > 0, \quad i = 1, 2, \ldots N,$$

is continuous and bounded on (0,a) and for any x, it must have a minimum (or maximum).

Let

$$f_N(x) = \min_{(x_i)} \left(h_1(x_1) + h_2(x_2) + \ldots . h_N(x_N) \right),$$

to show the implicit dependence on both x and N.
We note a subtle distinction here. The   sought
for   unknown   knots are the end of the intervals
consistent with the choice of x and N. In   two
cases the sequence $(f_N(x))$ takes on particularly
simple forms,

$$f_N(0) = 0, \quad \text{for all } N = 1, 2, \ldots$$

provided $h_i(0) = 0$ for $i = 1, 2, \ldots N$, and

$$f_1(x) = h_1(x).$$

We   wish now   to   obtain   a   recursive
relation between $f_N(x)$ and $f_{N-1}(x)$ and to
establish this we proceed as follows. If we
select x, $0 \leq x \leq x_N$, then regardless   of
the precise value of x_N, the remaining $x - x_N$
must be used to find a minimum from N-1
functions. Therefore, we can write directly,

$$f_N(x) = \min \ (h_N(x_N) + f_{N-1}(x-x_N)),$$

$$0 \leq x_N \leq x,$$

which is the Fundamental   Equation of Dynamic
Programming.

2.6   THE PRINCIPLE OF OPTIMALITY

Our foregoing results can be stated in the following way.

Principle of Optimality

On optimal value of $f_N(x)$ has the property that whatever the initial state, $f_1(x)$ and initial choice of x_1, the remaining choices must be optimal with regard to the state resulting from the first choice.

2.7   A DIRECT DERIVATION

The principle of optimality can be directly derived by observing.

$$\min_{x_1+x_2+\ldots+x_N=x} = \min_{0\leq x_N\leq x} \left(\min_{x_1+x_2+\ldots+x_{N-1}} \right)$$

$$x_i > 0, \qquad =(x-x_N)$$

so we can write

$$f_N(x) = \min (h(x_1) + h(x_2) + \ldots + h(x_N))$$
$$(x_1 + x_2 + \ldots + x_N = x)$$

$$x_i > 0,$$

$$= \min_{\substack{0 \leq x_N \leq x}} \left(\min \left(h_N (x_N) + h_{N-1} (x_{N-1}) \right) .. \right),$$

$$(x_1 + x_2 + ... + x_{N-1}) = (x - x_N)$$

$$= \min_{\substack{0 \leq x_N \leq x}} \left(h_N (x_N) + \min \left(h_{N-1} (x_{N-1} + ... \right), \right.$$

$$, x_1 + x_2 + .. x_{N-1} = (x - x_N)$$

$$= \min_{\substack{0 \leq x_N \leq x_N}} \left(h_N (x_N) + f_{N-1} (x - x_N) \right),$$

which is the Fundamental Equation of Dynamic Programming. In the following section we will use this result to obtain a linear approximation to f(x).

2.8   CURVE FITTING BY SEGMENTED STRAIGHT LINES

Consider the function g(x), whose values are given on the interval (0,a).   We wish to approximate this curve by a sequence of straight lines segments, each connecting it neighbor at the end points, see Figure 2.1. The problem is to determine the end points of each segment in such a way that the final curve best approximates g(x) over the entire interval (0,a).
We observe that if any two points in the domain of definition of g(x) have the coordinates (s_{ij}, x_k) and (s_{kl}, x_l), then a straight line connecting these points can be written as,

$$y(x) = \left(\frac{s_{ij} - s_{kl}}{x_i - x_k} \right) (x - x_k) + s_{kl} .$$

Let the measure of the error between the given curve g(x) and its straight line approximation

$y(x)$ in the interval (x_i, x_k) be simply defined as,

$$D(s_{ij}, s_{kl}) = \max_{x_k \leq x \leq x_i} \left| g(x) - \left(\left(\frac{s_{ij} - s_{kl}}{x_i - x_k} \right)(x - x_k) + s_{kl} \right) \right|$$

(2-8)

The problem is to determine the end points which, when connected, gives the minimum error as defined in Eq.(2-8).

2.9   A DYNAMIC PROGRAMMING APPROACH

In the spirit of dynamic programming, let the functional $f_N(b,a)$ be given by N segmented straight lines when the right end point of the of the last straight line segment is located at the point (a,b).Then for $N > 1$,the principle of obtimality yields the functional equation,

$$f_N(s_{ij}, x_i) = \min_{(s_{kl}: x_1 < x_k < x_i)} (D(s_{ij}, s_{kl}) + F_{N-1}(s_{kl}, x_k))$$

$$f_1(s_{ij}, x_i) = \min_{s_{11}} D(s_{ij}, s_{11}).$$

The computational problem is to determine $f_N(b,a)$ and the solution is determined by that value of b for which $f_N(b,a)$ is a minimum.

A by-product of this solution will be the
set of end points which will determine the best
sequence of straight line segments approximating the curve g(x) on the interval (0,a).

2.10   A COMPUTATIONAL PROCEDURE

Computationally, the evaluation of $f_N(b,a)$
can be found easily and the procedure can be
computerized. In general we will not allow x
to range over the entire interval nor will we
in general take on all values between the
minimum and the maximum values of g(x).

Rather we introduce a grid system into
the problem. Let the set determine the grid
points along the ordinate. Then the points x_i
and x will lie on the abscissa grid and the
points s_{ij}, s_{kl} will lie on the ordinate grid.
For the proper evaluation of the expression
$D(s_{ij}, s_{kl})$, we introduce a second finer grid
spacing Δx_2 which will define a finer mesh
within the interval (x_i, x_k). With these quantities defined, the problem can be easily solved
either by hand or by computer.

2.11   THREE DIMENSIONAL POLYGONAL APPROXIMATION

A completely different problem arises in the modern methods of approximation when one considers approximating a continuous function $f(x,y,z)$ within a bounded volume by a set of polynomials. The results of this section of the chapter will be used later when we discuss finite elements, a very successful modern method of approximation.

Let us consider the continuous function $f(x,y,z)$ defined in a volume V having a surface S. We wish to approximate the function $f(x,y,z)$ by the sum of quadratic polynomials in each of three independent directions. We begin by defining a set of 27 shape functions defined on the unit cube, $-1 \leq \xi,\eta,\zeta \leq 1$ shown in Figure 2.2.

Figure 2.2

The Finite Element Unit Cube

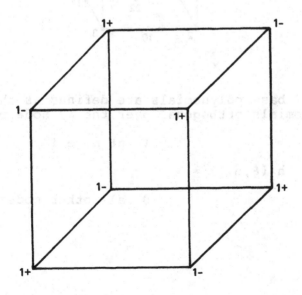

Since any quadratic polynomial is made up of
a linear combination of three basic functions,1, x,and x squared, the full three dimensional approximation requires 27 linearly independent functions. We seek to identify these 27
polynomials in a convenient manner by introducing 27 node points within the unit cube,as
shown in Figure 2.3.

Figure 2.3

The Finite Element Node Numbering

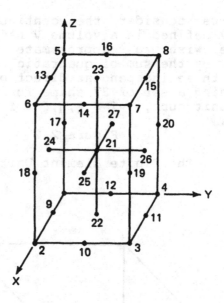

The 27 base polynomials are defined as the set of
polynomials orthogonal over the 27 node points.

1   at node i

$$h_i(\xi,\eta,\zeta) =$$

0   all other nodes

2.12 THE ORTHOGONAL POLYNOMIALS

The set of 27 orthogonal polynomials are,

$$h_1 = -\xi*(1-\xi)*\eta*(1-\xi)\zeta*(1-\zeta)/8$$

$$h_2 = \xi*(1+\xi)*\xi*(1-\xi)*\zeta*(1-\zeta)/8$$

$$h_3 = -\xi*(1+\xi)*\eta*(1+\eta)*\zeta*(1-\zeta)/8$$

$$h_4 = \xi*(1-\xi)*\eta*(1+\eta)*\zeta*(1-\zeta)/8$$

$$h_5 = \xi*(1-\xi)*\eta*(1-\eta)*\zeta*(1+\zeta)/8$$

$$h_6 = -\xi*(1+\xi)*\eta*(1-\eta)*\zeta*(1+\zeta)/8$$

$$h_7 = \xi*(1+\xi)*\eta*(1+\eta)*\zeta*(1+\zeta)/8$$

$$h_8 = -\xi*(1-\xi)*\eta*(1+\eta)*\zeta*(1+\zeta)/8$$

$$h_9 = (1-\xi**2)*\eta*(1-\eta)*\zeta*(1-\zeta)/4$$

$$h_{10} = -\xi*(1+\xi)*(1-\eta**2)*\zeta*(1-\zeta)/4$$

$$h_{11} = -(1-\xi**2)*\eta*(1+\eta)*\zeta*(1-\zeta)/4$$

$$h_{12} = \xi*(1-\xi)*(1-\eta**2)*\zeta*(1-\zeta)/4$$

$$h_{13} = -(1-\xi**2)*\eta*(1-\eta)*\zeta*(1+\zeta)/4 \quad (2-9)$$

$$h_{14} = \xi*(1+\xi)*(1-\eta**2)*\zeta*(1+\zeta)/4$$

$$h_{15} = (1-\xi**2)*\eta*(1+\xi)*\zeta*(1+\zeta)/4$$

$$h_{16} = -\xi*(1-\xi)*(1-\eta**2)*\zeta*(1+\zeta)/4$$

$$h_{17} = \xi*(1-\xi)*\eta*(1-\xi)*(1-\zeta**2)/4$$

$$h_{18} = -\xi*(1+\xi)*\eta*(1-\xi)*(1-\zeta**2)/4$$

$$h_{19} = \xi*(1+\xi)*\eta*(1+\eta)*(1-\zeta**2)/4$$

$$h_{20} = -\xi*(1-\xi)*\eta*(1+\eta)*(1-\zeta**2)/4$$

$$h_{21} = (1-\xi**2)*(1-\eta**2)*(1-\zeta**2)$$

$$h_{22} = -(1-\xi**2)*(1-\eta**2)*\zeta*(1-\zeta)/2$$

$$h_{23} = (1-\xi**2)*(1-\eta**2)*\zeta*(1+\zeta)/2$$

$$h_{24} = -(1-\xi**2)*\eta*(1-\eta)*(1-\zeta**2)/2$$

$$h_{25} = \xi*(1+\xi)*(1-\eta**2)*(1-\zeta**2)/2$$

$$h_{26} = (1-\xi**2)*\eta*(1+\eta)*(1-\zeta**2)/2$$

$$h_{27} = -\xi*(1-\xi)*(1-\eta**2)*(1-\zeta**2)/2$$

2.13   THE APPROXIMATION TECHNIQUE

Often we are confronted with problems involving integrals of unknown functions over a volume V. Applications immediately come to mind in variational calculus where we are required to find the function f(x,y,z) for which,

$$\int_V F(f) \; dv,$$

is a minimum, for the convex functional F(f).

We may approximate f(x,y,z) over the volume V by the expression,

$$f(x,y,z) = \sum_i f_i \; h_i(\xi,\eta,\zeta), \qquad (2\text{-}10)$$

where f_i is the value of the function at the space point (x_i, y_i, z_i) i = 1 , ... , 27 . In addition, the coordinates of all points within the volume V can be characterized by the mapping,

$$x = \sum_i x_i \, h_i(\xi,\eta,\zeta),$$
$$y = \sum_i y_i \, h_i(\xi,\eta,\zeta), \hspace{2cm} (2\text{-}11)$$
$$z = \sum_i z_i \, h_i(\xi,\eta,\zeta).$$

Expressions Eq.(2-10) and Eq.(2-11) represent a continuous mapping from the unit cube ($-1 \leq \xi,\eta,\zeta \leq 1$) to the points in V and the value of the function within the same volume. To use this approximation for interpolation, f must be known and Eq.(2-11) must be inverted, which is quite difficult. However, the evaluation of the integrals are straightforward.

$$\Psi = \int_V F(f) \, dv = \int_V F(\sum_i f_i \, h_i(\xi,\eta,\zeta)) \, |J| \, d\xi d\eta d\zeta,$$

where $|J|$ is the Jacobean of the transformation Eq.(2-11).

By using the 27 node approximation of a function $f(x,y,z)$ over a volume V we obtain a very accurate evaluation of the integral in terms of the 27 parametric nodes points. This result will be fully exploited in a later chapter on finite elements.

2.14 DISCUSSION

Let us finally note that in trying to approximate the continuous function $f(x)$ by a sequence of segmented straight lines, we are assuming something about the basic structure of the problem, mainly that the system we are approximating is linear (or near linear) between a finite but unknown set of critical states. Such problems must be chosen with great care, but having done so, we have at our disposal the full power of dynamic programming.

On the other hand, the use of the full 27 polynomial set of quadratics in the three dimensional approximation of f(x,y,z) over the volume V doesn't impose the same assumed structure of the problem as in the former case. The 27 polynomial set forms a linear independent set of polynomials spanning the set of all quadratic functions in three dimensions. The assumption involved in this approximation is the variation of f(x,y,z) within the volume on integration be not more than a quadratic. Of course, if the volume is taken small enough, the assumption is very nearly true, making the finite element method of approximation very useful.

2.15 BIBLIOGRAPHY AND COMMENTS

In 2.1, the Weierstrass Polynomial Approximation Theorem together with the algebraic properties of polynomials are found in

Gillman,L. and M.Jerison, :1960, Rings of Continuous Functions Springer

Titchmarsh,E.C.:1939, Theory of Functions, Oxford University Press

In 2.3, Curve fitting by straight lines is discussed in,

Bellman,R. and R.S.Roth,:1969, "Curve Fitting by Segmented Straight Lines", J.Amer. Stat.Assoc, 64, 1079-1084

Basic studies in Dynamic Programming are found in,

Bellman,R.:1957, Dynamic Programming, Princeton University Press,Princeton, N.J.

Bellman,R. and S.Dreyfus,:1962, Applied Dynamic Programming, Princeton University Press, Princeton,N.J.

For a treatment of the finite element method which arises in the three dimensional polygonal approximation, see,

Gallagher,R.H.:1975, Finite Element Analysis Fundamentals ,Prentice-Hall Englewood,N.J.

Zienkiewicz,O.C.:1971, The Finite Element Method in Engineering Science, McGraw-Hill, London

Chapter 3

POLYNOMIAL SPLINES

3.1 INTRODUCTION

In the last chapter we considered two very simple polynomials as approximating functions, the segmented straight line over the closed interval $(0,1)$, and the set of 27 orthogonal quadratic polynomials over the unit cube, $-1 \leq \xi, \eta, \zeta \leq 1$.

Now we wish to consider the polynomial spline as a modern approximation tool. In particular, we will focus on the cubic spline to illustrate the use of these special polynomials in approximation theory.

3.2 THE CUBIC SPLINE I

Let $f(x)$ be a continuous, bounded function defined in the closed interval $0 \leq x \leq 1$. We subdivide the interval by a set of mesh points, called nodes, and denote the set by Δ:

$$\Delta: \quad 0 = x_0 < x_1 < \ldots < x_N = 1.$$

Let

$$y_i = f(x_i), \quad i = 1, \ldots N.$$

A cubic spline $s\Delta(x)$ is a continuous polynomial function for $0 \leq x \leq 1$ such that,

(1) $s\Delta(x)$ is a cubic polynomial in every interval

$$(x_{j+1}, x_j),$$

(2) $s'\Delta(x)$ is continuous on $0 \leq x \leq 1$,

(3) $y_i = s\Delta(x_i)$, $i = 1, 2, \ldots N$.

3.3   CONSTRUCTION OF THE CUBIC SPLINE

The cubic spline can be constructed from the fact that within an interval the second derivative of the spline is linear.

By designating M_j as the moment of the spline $s\Delta(x)$ at x_j, we may say that within the interval (x_{i-1}, x_i),

$$s''\Delta(x) = M_{j-1}\left(\frac{x_j - x}{h_j}\right) + M_j\left(\frac{x - x_{j-1}}{h_j}\right),$$

$$(3-1)$$

where $h_j = x_j - x_{j-1}$. If we integrate Eq.(3-1) twice and evaluate the constants of integration, we have

$$s\Delta(x) = M_{j-1}\left(\frac{(x_j - x)^3}{6h_j}\right) + M_j\left(\frac{(x - x_{j-1})^3}{6h_j}\right)$$

$$+ \left(y_{j-1} - \frac{M_{j-1}h_j^2}{6}\right)\left(\frac{x_j - x}{h_j}\right)$$

$$\text{(3-2)}$$

$$+ \left(y_j - \frac{M_j h_j^2}{6}\right)\left(\frac{x - x_j}{h_j}\right),$$

and

$$s'\Delta(x) = -M_{j-1}\left(\frac{(x_j - x)^2}{2h_j}\right) + M_j\left(\frac{(x - x_{j-1})^2}{2h_j}\right)$$

$$- \left(\frac{M_j - M_{j-1}}{6}\right)h_j$$

$$+ \left(\frac{y_j - y_{j-1}}{h_j}\right).$$

In expression Eq.(3-2), only the set of moments M_j, $j = 0$, ... N, are unknown and must be determined. Since the spline

requires continuity in the first derivative
at each node we can write

$$s'\Delta(x_j^-) = \frac{h_j}{6} M_{j-1} \div \frac{h_j}{3} M_j + \left(\frac{y_j - y_{j-1}}{h_j}\right),$$

$$(3-3)$$

$$s'\Delta(x_j^+) = - \frac{h_{j+1}}{3} M_j - \frac{h_{j+1}}{6} M_{j+1}$$

$$(3-4)$$

$$+ \left(\frac{y_{j+1} - y_j}{h_{j+1}}\right),$$

defining the right and left derivatives of the
cubic spline at x_j.

Equating Eq.(3-3) and Eq.(3-4) at each
node x_j to ensure continuity, we have

$$\frac{h_j}{6} M_{j-1} + \frac{(h_j + h_{j+1})}{3} M_j + \frac{h_j}{6} M_{j+1} =$$

$$\left(\frac{y_{j+1} - y_j}{h_{j+1}}\right) - \left(\frac{y_j - y_{j-1}}{h_j}\right).$$

Two additional conditions must be satisfied to determine the N+1 unknowns $M_0, M_1 \ldots M_N$ A common technique is to specify the slopes at the end points. If $s'\Delta_0 = y'_0$, $s'\Delta_N = y'_N$, then using Eq.(3-3) and Eq.(3-4),

$$2 M_0 + M_1 = \frac{6}{h_1} \left(\frac{y_1 - y_0}{h_1} - y_0' \right),$$

$$M_{N-1} + 2 M_N = \frac{6}{h_N} \left(y_N' - \frac{y_N - y_{N-1}}{h_N} \right).$$

We can also put $M_0 = M_N = 0$ or $M_0 + \lambda M_1 = 0$, $0 < \lambda < 1$.

We will explore such ideas further later in the chapter. In general we can write the spline end conditions as,

$$2 M_0 + \lambda(0) M_1 = d_0$$

$$\mu(N) M_{N-1} + 2 M_N = d_N.$$

If we let

$$\lambda(j) = \frac{h_{j+1}}{h_{j+1} + h_j} \quad , \quad \mu(j) = 1 - \lambda(j)$$

then the continuity condition can be written,

$$\mu(j) M_{j-1} + 2 M_j + \lambda(j) M_{j+1} = 6 d_j,$$

$$j = 1, 2, \ldots N-1, \qquad (3-4)$$

where

$$d_j = \frac{\left(\dfrac{y_{j+1} - y_j}{h_{j+1}} \right) - \left(\dfrac{y_j - y_{j-1}}{h_j} \right)}{(h_j + h_{j+1})}.$$

The entire system can then be written,

$$
\begin{vmatrix}
2 & \lambda(0) & \cdots & 0 & 0 \\
\mu(1) & 2 & \cdots & 0 & 0 \\
0 & \mu(2) & & \cdot & \\
 & & & \cdot & \\
 & & & \cdot & \\
 & & 2 & \lambda(N-1) & \\
 & & \mu(N) & 2 &
\end{vmatrix}
\begin{vmatrix}
M_0 \\
M_1 \\
M_2 \\
\cdot \\
M_{N-1} \\
M_N
\end{vmatrix}
=
\begin{vmatrix}
d0 \\
d_1 \\
d_2 \\
\cdot \\
d_{N-1} \\
d_N
\end{vmatrix}.
$$

3.4   EXISTENCE AND UNIQUENESS

We begin this section with a review of the Gersgorin circle theorem. Let λ be an eigenvalue of the matrix A and x the corresponding eigenvector. Therefore we can write,

$$A x = \lambda x$$

let x_m be the maximum nonzero eigenvector, then

$$\sum_j a_{mj} x_j = \lambda x_m$$

or by extracting the pivot element,

$$\lambda - a_{mm} = \sum_{j \neq m} a_{mj} \frac{x_j}{x_m}$$

we obtain the inequality,

$$| \lambda - a_{mm} | \leq \sum_{j \neq m} |a_{mj}|. \qquad (3\text{-}5)$$

Eq.(3-5) mearly states that the eigenvalues λ lie within a circle in the complex plane centered at a_{mm} with a radius $\sum |a_{mj}|$. Therefore we have the Gersgoren theorem.

The eigenvalues of a matrix A lie in the union union of the circles,

$$|z - a_{ii}| \leq \sum_{j \neq i} |a_{ij}| \qquad (3\text{-}6)$$

$$0 < i \leq n.$$

We observe that if the matrix A has no zero eigenvalues,then the matrix is nonsingular and a unique inverse exists.

Now if we define a dominant main diagonal matrix A by the property,

$$|a_{ii}| \geq \sum_{j \neq i} |a_{ij}|,$$

then by Eq.(3-6),

$$|z - a_{ii}| < |a_{ii}|. \qquad (3-7)$$

To satisfy Eq.(3-7), z must be nonzero and we can say that a dominate main diagonal matrix is nonsingular.

Now we know that the success of the cubic spline depends on the nonsingularity of the matrix,

$$
\begin{vmatrix}
2 & \mu(0) & & & & \\
\lambda(1) & 2 & \mu(1) & & & \\
& & & \ddots & & \\
& & & \mu(N-1) & 2 & \lambda(N-1) \\
& & & & \mu(N) & 2
\end{vmatrix}
$$

which is nonsingular since it is clearly a dominant main diagonal matrix as seen from the definition of λ and μ.

3.5   A COMPUTATIONAL ALGORITHM - POTTER'S METHOD

The system of spline equations defined in the last section can be written in the form,

$$2 M_0 + M_1 = d_0 ,$$

$$\mu(j) M_{j-1} + 2 M_j + \lambda(j) M_{j+1} = d_j ,$$

$$j = 1,2,\ldots N-1,$$

(3-8)

$$M_{N-1} + 2 M_N = d_N .$$

A matrix inversion will solve the system, but since the matrix has elements centered on the diagonal is a neater way to do it. Let

$$M_j = a(j) M_{j+1} + b(j), \qquad (3-9)$$

where for the moment $a(j)$ and $b(j)$ are unknown. Substituting Eq.(3-9) into Eq.(3-8) we have

$$\mu(j)(a(j-1)M_j + b(j-1)) + 2 M_j + \lambda(j)M_{j+1} = d_j,$$

or, by rearranging terms

$$M_j = \frac{-\lambda(j)}{(2 + \mu(j) a(j-1))} M_{j+1}$$

(3-10)

$$+ \quad \frac{(d_j - \mu(i)\, b(j-1))}{(2 + \mu(j)\, a(j-1))} \; .$$

Comparing Eq.(3-9) with Eq.(3-10) leads to the recursion relation,

$$a(j) = \frac{-\lambda(j)}{(2 + \mu(j)\, a(j-1))} \; , \qquad (3-11)$$

and

$$b(j) = \frac{(d(j) - \mu(j)\, b(j-1))}{(2 + \mu(j)\, a(j-1))} \; ,$$

$$j = 1,2 \ldots N-1.$$

Once a(0) and b(0) are established, a(j) and b(j) can be computed for j = 1,..N-1 using Eqs.(3-11). The initial conditions for a(0) and b(0) are found in terms of the end conditions which are given as follows.

The end conditions can be expressed in a different form,
at $x = x_0$,

$$2 M_0 + M_1 = \left(\frac{6}{h_1}\right)\left(\frac{y_1 - y_0}{h_1} - y'_0\right),$$

$$M(0) = 2y''(0)$$

$$(3-12)$$

$$2M_0 + \lambda(0) M_1 = d_0$$

while at $x = x_N$,

$$M_{N-1} + 2 M_N = \left(\frac{6}{h_N}\right) \left(y'_N - \frac{y_N - y_{N-1}}{h_N} \right),$$

$$2 M_N = 2 y''_N ,$$

$$\mu(N) M_{N-1} + 2 M_N = d_N .$$

At the x_0 end point the initial values of $a(0)$ and $b(0)$ are known for since

$$M_j = a(j) M_{j+1} + b(j) .$$

If the initial slope is known from observed data , $y'_0 = (y_1 - y_0)/h_1$, then, from Eq.(3-12),

$$M_0 = - (1/2) M_1 ,$$

and comparing this result with Eq.(3-9) for $j = 0$, gives,

$$a(0) = - (1/2), b(0) = 0.$$

On the other hand, if y''_0 is known, then $a(0)=0$, $b(0) = 2y''_0$ and finally, if d_0 is known, then again by Eq.(3-9) we have,

$$M_0 = -\lambda(0)/2 \, M_1 + d_0/2,$$

giving,

$$a(0) = -\lambda(0)/2, \quad b(0) = d_0/2.$$

With this information, following the recursive relation Eq.(3-11), we can compute and store all coefficients $a(i), b(i), i = 1, 2, \ldots, N-1$.

When $j = N-1$, we have the relations,

$$M_{N-1} = a(N-1) \, M_N + b(N-1),$$

$$\mu(N) \, M_{N-1} + 2 \, M_N = d_N.$$

Solving for M_N,

$$M_N = \frac{d_N - \mu(N) \, b_{N-1}}{2 + \mu(N) \, a(N-1)}.$$

Finally, since $M_i = a(i) \, M_{i+1} + b(i)$, all the spline moments can be computed.

The procedure we have just described is interesting in its own right, for it describes, in mathematical terms, the propagation of partial information. The partial information represented by the end conditions is propagated thru the parameters $a(i)$, $b(i)$ until the end is reached. Knowledge at that point allows us to compute the first (actually last) spline moment The recursion relation then allows us to compute all remaining moments. This property is typical of the Riccati equation which we shall consider in a later chapter.

The spline polynomial, defined over the interval (x_{j-1}, x_j) can be written in terms of the spline moments.

$$s_j(x) = \left(\frac{M_j}{6h_j} - \frac{M_{j-1}}{6h_j} \right) x^3$$

$$+ \left(x_j \frac{M_{j-1}}{2h_j} - x_{j-1} \frac{M_j}{2h_j} \right) x^2$$

$$+ \left(-x_j^2 \frac{M_{j-1}}{2h_j} + x_{j-1} \frac{M_j}{2h_j} \right)$$

$$- \left(\frac{y_{j-1} - M_{j-1} h_j^2 /6}{h_j} \right))$$

$$+ \left(\frac{y_j - M_j h_j^2 /6}{h_j} \right) x$$

$$+ \left(x_j^3 \frac{M_{j-1}}{6h_j} - x_{j-1}^3 \frac{M_j}{6h_j} \right)$$

$$+ \left(\frac{y_{j-1}}{h_j} - \frac{M_{j-1} h_j}{6} \right) x_j$$

$$- \left(\frac{y_j}{h_j} - \frac{M_j h_j}{6} \right) x_{j-1}$$

3.6   SPLINES VIA DYNAMIC PROGRAMMING

The method of splines approximates a function in a piecewise fashion by means of a different polynomial in each subinterval. Having its origin in beam theory, the spline processes continuous first and second derivatives at the nodes and is characterized by a minimum curvature property which is related to the physical feature of minimum potential energy of the supported beam.

Translated into mathematical   terms this means that between successive nodes the approximation yields a third order polynomial with its first two derivatives continuous at ' the node points. The minimum curvature property holds good for each subinterval as well as the whole region of approximation.This means that the integral of the square of the second

derivative over the entire interval as well as
each subinterval is to be minimized. Then,the
task of determining the spline offers itself as
a textbook problem in discrete dynamic programming since the integral of the square of the
second derivative can be recognized as the criterion function which must be minimized. Starting with the initial value of the function and
assuming an initial slope of the curve, the
minimum property of the curvature makes se-
quential decision of the slope at successive
node points feasible.

3.7   DERIVATION OF SPLINES BY DYNAMIC
         PROGRAMMING

We subdivide the interval $a < x \leq b$, into
N node points so,

$$\Delta : \quad a = x_0 \quad < x_1 \quad < \ldots < x_N \quad = b.$$

Let $f(x)$ be defined on the interval with the
prescribed values,

$$Y: \quad y_i = f(x_i), \quad i = 0,1,\ldots N.$$

We seek to approximate $f(x)$ by $s(x)$ which is continuous with its first and second derivatives
in (a,b). Consider a cubic spline in each interval
$x_{j-1} < x \leq x_j$, $j = 1,2,\ldots N$ and satisfies the
condition $s(x_i) = y(x)$. Holladay's theorem states
that the spline function $s(x)$ with $s''(a) = s''(b)$
$= 0$, minimizes the integral,

$$J(a,b) = \int_a^b (f''(x))^2 \, dx.$$

We can now apply the methods of dynamic programming to determine $s(x)$ in each sub-interval.   Let us consider the interval $x_i \leq x \leq x_{i+1}$ and write,

$$s_i(x) = y_i + M_i(x - x_i) + p(i)(x-x_i)^2$$

$$+ g(i)(x - x_i)^3,$$

for $x_i \leq x \leq x_{i+1}$.

Since y_i and y_{i+1} are known, letting M_i and M_{i+1} denote the first moments at the nodes x_i and x_{i+1}, we find, using the continuity property, that,

$$p(i) = (\frac{1}{h_{i+1}})(3r(i) - (M_{i+1} + 2 M_i),$$

$$q(i) = (\frac{1}{h_{i+1}^2}) ((M_{i+1} + M_i) - 2 r(i)),$$

where

$$r(i) = (y_{i+1} - y_i)/h_{i+1},$$

$$h_{i+1} = (x_{i+1} - x_i).$$

In order to arrive at the multistage decision process for the determination of the moments at successive nodes, we write the minimum
value of the integral corresponding to the minimum curvature property, starting with the node
point,

$$F_i(M_i) = \min_{\substack{M_{i+k} \\ k=1,N-1}} \sum_{j=i}^{N-1} \int_{x_j}^{x_{j+1}} (s''(x))^2 \, ds,$$

(3-13)

$$x_N = b.$$

Here we have assumed that the first moment at x_i
is M_i. All the subsequent values of the slopes
M_{i+k}, $(k > 1)$ are to be found in terms of
M_i by the minimization condition. Eq.(3-13),in
the equal fashion leads to the recursion relation-
ship,

$$F_i(M_i) = \min_{M_{i+1}} (\int_{x_i}^{x_{i+1}} (s''(x))^2 \, dx + F_{i+1}(M_{i+1}),$$

$$= \min_{M_{i+1}} (g(i+1)(3r(i)^2 + M_{i+1}^2 + M_i^2$$

(3-14)

$$-3r(i)(M_{i+1} + M_i) + M_i M_{i+1})$$

$$+ F_{i+1}(M_{i+1})),$$

where $g(i+1) = 4/h_{i+1}$.

We now argue that F_i is quadratic in M_i. To see this, let us take the last interval, $x_{N-1} \leq x \leq X_N$,

$$F_{N-1}(M_{N-1}) = g(N)(3/4\ r(N-1)^2 - 3/2\ r(N-1)M_{N-1} \qquad (3\text{-}15)$$

$$+ 3/4\ M_{N-1}^2 .$$

The above expression is clearly quadratic in M_{N-1}. Hence, by induction, it can be shown that F_i is quadratic in M_i. Therefore write,

$$F_i(M_i) = c(i) + d(i)M_i + e(i)\ M_i^2 . \qquad (3\text{-}16)$$

From Eq.(3-14) we obtain,

$$F_i(M_i) = \operatorname*{Min}_{M_{i+1}} (g(i+1)(3r(i)^2 + M_{i+1}^2 + M_i^2$$

$$(3\text{-}17)$$

$$-3r(i)(M_{i+1} + M_i) + M_i M_{i+1}))$$

$$c(i+1) + d(i+1)M_{i+1} + e(i+1)M_{i+1}^2).$$

The minimization over M_{i+1} is readily performed.

Thus   Eq.(3-14) and   Eq.(3-17) lead to the following recursion relation,

$$M_{i+1} = \frac{3r(i)g(i+1) - g(i+1)M_i - d_{i+1}}{2 \ (g(i+1) + e(i+1))} \ ,$$

$$c(i) = c(i+1) + 3r(i)^2 g(i+1) - \frac{(3r(i)g(i+1)-d_{i+1})^2}{4(g(i+1) + e(i+1))} \ ,$$

$$\text{(3-18)}$$

$$e(i) = g(i+1)\left(1 - \frac{g(i+1)}{4(g(i+1)+e(i+1))} \right),$$

$$d(i) = g(i+1)\left(\frac{3r(i)g(i+1) - d_{i+1}}{2(g(i+1)+e(i+1))} - 3r(i)\right).$$

From Eq.(3-15) the values of $c(N-1), d(N-1)$ and $e(N-1)$ are given by

$$c(N-1) = 3r(N-1)^2/h_N \ ,$$

$$d(N-1) = -6 \ r(N-1)/h_N \ ,$$

$$e(N-1) = \ 3/h_N \ .$$

Hence $c(i), d(i)$ and $e(i)$ for $i = 0,1,\ldots N-1$ can be easily determined from the recursive relations.Similarly,all the moments can be obtained in   terms of the assumed value   M of the slope at $x = a$.

The minimum value assumes the form

$$F_0(M_0) = c(0) + d(0) M_0 + e(0) M_0^2.$$

The value of M_0 which minimizes this expression is the starting value of the recursion equation for the M's. Once the quantities are known, the spline function over each interval is known.

3.8 EQUIVALENCE OF THE RECURSIVE RELATIONS OBTAINED BY DYNAMIC PROGRAMMING AND THE USUAL RESULTS

Let us consider the recursive relations and the coefficients $c(i), d(i)$ and $e(i)$ in Sec. 3.7. We can easily rewrite the recursion re-lations as,

$$M_{i+1} = \frac{d_i}{g(i+1)} + 3r(i) - \frac{g(i+1)M_i}{2(g(i+1) + e(i+1))},$$

and

$$2(g(i) + e(i)) M_i = 3r(i-1)g(i) - d(i) - g(i)M_{i-1}.$$

Adding these two relations and using Eq.(3-17) we have,

$$\lambda(i)M_{i-1} + 2 M_i + \mu(i)M_{i+1} = 3\lambda(i) \frac{(y_i - y_{i-1})}{h_i}$$

$$+3\mu(i) \ \frac{(y_{i+1} - y_i)}{h_{i+1}} \ ,$$

where $\lambda(i) = h_{i+1}/(h_i + h_{i+1})$, and
$\mu(i) = 1 - \lambda(i)$.

This is precisely the recursion relation obtained by the usual method. The end relations in our case is given by the minimization procedure we adapted. Thus at x=b, minimizing $F_{N-1}(y_{N-1})$ yields the relation,

$$2 M_N + M_{N-1} = \frac{3(y_N - y_{N-1})}{h_N} \ .$$

At x = a minimizing $F_0(M_0)$ yields

$$M_0 = -\frac{d_0}{2 \ e(0)} \ . \qquad (3\text{-}19)$$

We know that,

$$M_1 = \frac{3r(0)g(1) - g(1)M_0 - d_1}{2(g(1) + e(1))} \ ,$$

$$(3\text{-}20)$$

where $r(0) = (y_1 - y_0)/h_1$ and $g(1) = 4/h_1$,

with h $= x - x$.
     1   1   0

Making use of the relation obtained earlier for d and e and using equations Eq.(3-19) and Eq.(3-20), we have

$$M + 2 M = 3r(0) \text{ , at } x=a.$$
    1   0

These two end conditions express the fact that $p(0)$ and $p(N)$, the second moments at the node ends are zero.

Thus we have found that our recursive relations are equivalent to the usual results corresponding to the end conditions mentioned earlier. Any other end conditions can be taken care of by the dynamic programming procedure by altering the criterion function in a suitable manner making use of Lagrangian multiplier techniques.

3.9 CARDINAL SPLINES

To tackle two point boundary value problems in ordinary differential equations with boundary conditions,we use cardinal splines which facilitate the sort of equation in a simple manner by converting it into a simple matrix inversion problem. If, for example,the boundary conditions involve the first derivatives of the function and the value of the function only at the boundaries,we think of a set of N+3 independent splines forming a basis for all cubic splines on the mesh Δ; . They are defined as follows, let,

$$Ak(x_j) = \delta_{kj} \quad , \ j=0,1,\dots N$$

$$Ak'(x_i) = 0 \quad , \ i=0 \text{ and } N, k=0,1\dots N$$

$$Bk(x_j) = 0 \quad , \ j=0,1,\dots N$$

$$Bk'(x_i) = \delta_{ik} \quad , \ i=0 \text{ and } N, \ k=0 \text{ and } N.$$

Then the interpolating function is given by,

$$S(y,x) = \sum_{j=0}^{N} (Aj(x)y(x_j)+y'(a)B0(x)+y'(b)Bn(x)$$

$$+ y'(b) \ Bn(x).$$

We want to point out that the Ak and Bk splines can be computed by dynamic programming methods since the boundary conditions Ak' = c for all k can be incorporated into the S function by adding λk(Ak-c)(Ak-c) terms and carrying out the usual minimization procedure(see Sec. 3.7). The Lagrangian parameters are to be determined by minimization. The recursive relations for the coefficients Eq.(3-18) and the relations for M's can be calculated quite easily since for each Ak only two of the r(i)'s are nonzero. Similarly the Bk splines can be evaluated and hence S(y,x). Substituting the spline into the differential equation, and adding the end conditions,one arrives at a matrix equation for the function values and its derivatives at the boundaries.

3.10   POLYNOMIAL SPLINES

A natural extension of the cubic spline is
the odd order polynomial of degree (2n-1). The
even order polynomial spline has to be treated
in a different manner. If f(x) is an arbitrary
function in the interval (a,b) having n-1 derivatives at x=a and b, then the spline representation S(f) is such that

$$\int_a^b (S^{(N)}(f))^2 \, dx,$$

is a minimum when S satisfies the end conditions that all its derivatives vanish. This
extension to higher order polynomials offer no
difficulty in treating the problem using the
method of dynamic programming, for the 2N-1 order polynomial at each mesh point the value
of the function assumes the first(n-1)   derivatives there and determines the moments
in each interval. Thus the minimization has
to be   carried   over   all the   (N-1) assumed moments at each mesh point in order to
determine sequentially   the moments   at the
next point in the mesh.   If, for example, we
take the fifth order polynomial spline, we have
to assume at any point x the   first and second
moments and p(i)   respectively and express the
other coefficients in S(f) in terms of the values of the function and the first two derivatives at the points . Hence the integral,

$$\Psi = \int_{x_i}^{x_{i+1}} (S^{(3)}(x))^2 \, dx,$$

can be expressed in terms of the above quantities and the mesh interval between x_i and x_{i+1},

$$\Psi = C(y_i, M_i, p(i), y_{i+1}, M_{i+1}, p(i+1)).$$

The key equation of the dynamic programming method
for the case (analogous to Eq.(3-14) in the cubic
spline case) will be of the form,

$$F_i(M_i, p(i)) = \min_{(M_{i+1}, p(i+1))} (\Psi_{i+1} + F_{i+1}(M_{i+1}, p(i+1))).$$

Since Ψ is a known function of its arguments,
arguing as before that F is quadratic in
M_i and $p(i)$,   we can arrive at the recursive
relations for the other moments appearing in
the polynomial in each interval. Once F_0 has
been expressed in terms of the assumed values
of M_0 and $p(0)$, it must be minimized with respect
to both these variables again to obtain the
values of M_0 and $p(0)$. These in turn will determine
the moments in all the intervals as was seen in
the cubic spline case.

3.11   GENERALIZED SPLINES

The simple $(2N-1)$ order polynomial splines
are characterized by the minimum value   of the
integral,

$$\int_a^b (D^n S)^2 \, dx. \qquad (3-21)$$

where D is a differential operator, assuming that
the proper end conditions are assured   In fact,
the Euler equation corresponding to Eq.(3-20) is

$$D^{2n} S = 0.$$

Thus the (2N-1) order polynomial is the solution of the Euler equation. If,in the same vein,we generalize to the case in which D is replaced by a general linear differential operater L satisfying suitable conditions and if L* is the formal adjoint of such an operator,we find that S satisfied the differential equation,

$$L*LS = 0,$$

where $\int_a^b (LS)^2 dx$ is a minimum and suitable boundary conditions are satisfied.

If n is the order of the differential operator, the equation is of the order 2n. For each interval,the solution of this equation can be obtained in terms of the values of the function and (n+1) derivatives at both ends. Hence the value of the integral can be computed for each interval in terms of these quantities, one can proceed in the usual dynamic programming method. Assuming the values of the first (n-1) derivatives at any mesh point, the integral F can be expressed in terms of the (n-1) derivatives at i and suitable recursion relations for the coefficients in this function can be obtained as detailed earlier.

3.12 MEAN SQUARE SPLINE APPROXIMATION

Given a function f(t) over the interval(0,T) defined explicitly by means of a table of values,an analytical expression or implicitly by an equation,it is often desirable to obtain an approximation in terms of a function of a standard type. One way to accomplish this is by means of an interpolation formula for which the classical example is the Lagrange polynomial interpolation.

Given n+1 points, we seek a polynomial of
degree n, $P_n(x)$, with the property   that

$$P_n(t_i) = u(t_i), \quad i = 1,2,\ldots,n.$$

There are, however, many advantages to using, instead, a piecewise, or spline approximation.   Here we wish to discuss the use of dynamic programming to treat the corresponding
problem for splines restricting our attention
one again to the cubic spline.   The general
case can be treated by choosing the nodes in a
most convenient fashion, as a function of u(t)
can also be treated by means of dynamic programming.

3.13   THE CUBIC SPLINE II

The cubic spline, $Sn(x)$ with nodes (t_i) is
determined by the following conditions,

a) $Sn(t)$ is a cubic polynomial in each of the
intervals (t_k, t_{k+1}), k=0,1...N-1,
$$t_0 = 0, \quad t_N = T,$$

b) $Sn(t), Sn'(t), Sn''(t)$ are continuous throughout
(0,T).

It follows that $Sn(t)$ in the interval (t_k, t_{k+1}) is
completely specified once the values of $Sn(t)$ and
$Sn'(t)$ are specified at the end points. Furthermore as follows from the standard interpolation
formulas in (t_k, t_{k+1}) $Sn(t)$ is linear
in these values, ie.

$$Sn(t) = Sn(t_k)an(t) + Sn'(t_k)bn(t) + Sn(t_{k+1})cn(t)$$

$$+ Sn'(t_{k+1})dn(t),$$

where the polynomials $an(t), bn(t), cn(t)$ and $dn(t)$ depend on t and t_{k+1}.

3.14   THE MINIMIZATION PROCEDURE

Let us consider the family of minimization problems,

$$Min J_k(u) = min \int_{t_k}^{T} (u - Sn(t))^2 dt,$$

$$k=0,1,\ldots N-1.$$

We begin by considering the case when $Sn(t)$ and $Sn'(t)$ are specified at t_k. Then we will minimize over the choice of that value. We introduce the the function,

$$f_k(c_1,c_2) = min_R J_k(u),$$

$$k = 0, 1, \ldots N-1,$$

where R is the set of spline functions restricted by the conditions,

$$Sn(t_k) = c_1, \quad Sn'(t_k) = c_2.$$

The minimization over c_1 and c_2 will be readily accomplished since $f_k(c_1, c_2)$ is quadratic in c_1 and c_2 as follows inductively from the procedure below.

3.15   THE FUNCTIONAL EQUATION

Let us use the additive property of integrals,

$$\int_{t_k}^{T} = \int_{t_k}^{t_{k+1}} + \int_{t_{k+1}}^{T} .$$

The the principle of optimality in dynamic programming yields the relation,

$$f_k(c_1, c_2) = \min_{d_1, d_2} (q_k(c_1, c_2, d_1, d_2)$$

$$\qquad\qquad (3\text{-}22)$$

$$+ f_{(k+1)}(d_1, d_2)),$$

where $q_k(c_1, c_2, d_1, d_2)$ is the quadratic polynomial obtained from the evaluation of

$$\int_{t_k}^{t_{k+1}} (u - Sn(t))^2 \, dt ,$$

where $Sn(t)$ is the cubic polynomial determined by the conditions,

$$Sn(t_k) = c_1, \qquad Sn'(t_k) = c_2$$

$$Sn(t_{k+1}) = d_1, \qquad Sn'(t_{k+1}) = d_2.$$

3.16   RECURSION RELATIONS

The use of the quadratic character of f_k and q_k enables the minimization in Eq.(3-22) to be carried out explicitly and the recurrence relations for the coefficients of the function to be determined.

3.17   BIBLIOGRAPHY AND COMMENTS

In 3.2, polynomial splines are studied in

Alhberg,J.H.,E.N.Nilson and J.L.Walsh,:1967, The Theory of Splines and Their Applications, Academic Press,N.Y.

In 3.4, for a discussion of the Gersgorin Circle Theorem, see,

Gersgorin,S.:1941,"Uber die Abgrenzung die Eigenwerte einer Martix",Izv. Akad.Nauk SSSR. ser.Mat.,7,749-753

Todd,J.:1962, A Survey of Numerical Analysis McGraw-Hill Book Co, N.Y.

In 3.5, Potter's method is discussed in,

Potter,M.L.:1955," A Matrix Method for the

Solution of a Second Order Difference
Equation in two Variables"'
Math Centrum Report, MR19

In 3.7, for additional papers we note,

Bellman,R.,B.G.Kashef and R.Vasdevan,:1973,
"Splines via Dynamic Programming", JMAA,38,
2,427-430

Bellman,R.,B.G.Kashef and R. Vasudevan,:1973,
"A Note on Mean Square Spline Approximation",
JMAA 42,2 47-53

Bellman,R.,B.G.Kashef,R. Vasudevan, and E.S.
Lee,:1975 , "Differential Quadrature and
Splines", Computers and Mathematics with
 Applications, vol 1,3/4,372-376

Bellman,R. and R.S.Roth,:1971,"The Use of
Splines with Unknown End Points in the
Identification of Systems",JMAA,34,1,26-33

to investigate Holladay's theorem, see,

Holladay,J.C.:1957,"Smoothest Curve Approximation ", Math Tables Aids to Computation,11,
233-267

Chapter 4

QUASILINEARIZATION

4.1   INTRODUCTION

In this chapter we intend to explore several numerical techniques for fitting a known function to a linear or nonlinear differential equation. This technique, known as quasilinearization, makes specific use of the underlying structure of the linear differential equation allowing us to approximate, numerically, both initial conditions and system parameters associated with the selected differential equation.

If $u(t)$ is known on the interval $(0,1)$, we seek to determine a differential equation and boundary conditions b, having the form $W(v,b) = 0$ whose solution $v(t)$ best fits $u(t)$ over the interval according to some error criterion.

4.2   QUASILINEARIZATION I

There is no reason to believe that the best approximating differential operator is a linear equation with constant coefficients. On the other hand it must be clearly understood that if one seeks a differential approximation to u, $W(u,b)=0$, certain apriori knowledge of the form of $W(u,b)$ must be assumed.

If $W(u,b)$ is assumed to be nonlinear in u, then the approximation methods of the last chapter may have to be modified. Of course, if we proceed in exactly the same way as before,

we are lead to a set of nonlinear algebraic
equations which,in general, are quite difficult
to solve.   A very fruitful alternative is to
consider the method of quasilinearization which
is a generalization of the Newton Raphson method for extracting a root of the function f(x).

4.3   THE NEWTON RAPHSON METHOD

We shall begin with the problem of finding a
sequence of approximations to the root of the
scalar equation

$$f(x) = 0.$$

We shall assume $f(x)$ is monotone decreasing for
all x and strictly convex, $f''(x) > 0$, as shown
in Fig.   4.1.   Hence the root   is simple,
$f'(x) \neq 0$.

Figure 4.1

A Root of the Function f(x)

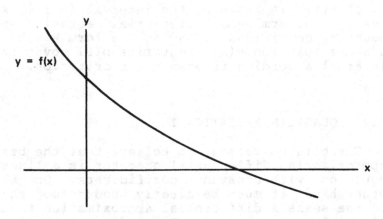

Suppose we have an initial approximation x_0 to
the root r   such that $x_0 < r$, $f(x_0) > 0$ and
let us expand $f(x)$ about x_0 in a Taylor's series,

$$f(x) = f(x_0) + (x - x_0)f'(x_0) + \ldots$$

A second approximation to r_0 can be obtained by solving the linear equation below for x,

$$f(x) = f(x_0) + (x - x_0) f'(x_0) .$$

This yields a second approximation,

$$x_1 = x_0 - \frac{f(x_0)}{f'(x_0)} .$$

If the process is repeated, we are lead to a recursion relation,

$$x_{n+1} = x_n - \frac{f(x_n)}{f'(x_n)} . \qquad (4-1)$$

We immediately note that Eq.(4-1) does not hold if $f'(x_n) = 0$ and because of the monotonicity of $f(x)$ ($f(x_n)>0$, $f(x_n)<0$) the sequence of values generated by Eq.(4-1) will be monotonically convergent,

$$x_0 < x_1 < x_2 < \ldots < x_n < r_0 .$$

as shown in Fig. 4.2.

Figure 4.2

Successive Approximation to the
Root of f(x)

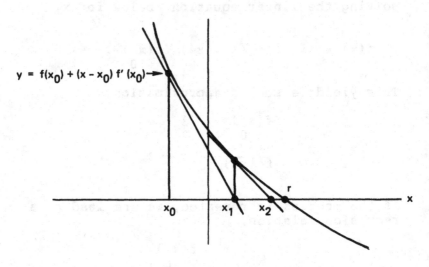

$y = f(x_0) + (x - x_0) f'(x_0) \rightarrow$

A second property of the Newton-Raphson method is that the monotone sequence converges quadratically, ie. the sequence has the property that,

$$|x_{n+1} - r_0| < k |x_n - r_0|^2 ,$$

where k is independent of n.

We consider the relation

$$r_0 - x_{n+1} = r_0 - x_n + \frac{f(x_n)}{f'(x_n)} .$$

Since $f(r_0) = 0$, we can write ,

$$r_0 - x_{n+1} = r_0 - x_n - \left(\frac{f(r_0) - f(x_n)}{f'(x_n)} \right).$$

(4-2)

By the mean value theorem,

$$f(r_0) - f(x_n) = (r_0 - x_n)f'(x_n)$$

$$+ \left(\frac{1}{2}\right)(r_0 - x_n)^2 f''(\xi),$$

(4-3)

where $x_n \leq \xi \leq r_0$.

Substituting Eq.(4-3) into Eq.(4-2) we have,

$$r0 - x_{n+1} = -\left(\frac{1}{2}\right)\left(\frac{f''(\xi)}{f'(x_n)}\right)(r_0 - x_n)^2.$$

If $f''(\xi), f'(x_n)$ exist and $f'(x_n) \neq 0$,
we set

$$k = \max_{x_0 \leq \xi \leq r_0} \left(\frac{1}{2}\right)\left(\frac{f''(\xi)}{f'(\xi)}\right).$$

Then

$$|r_0 - x_{n+1}| \leq k |r_0 - x_n|^2$$

and the convergence is quadratic.

4.4   QUASILINEARIZATION II

The quasilinear technique is a generalization of the Newton Raphson method developed in
the last section. As such we expect, and will
indeed show ,that the properties of monotonicity and quadratic convergence apply here as
well.

We shall begin by considering the nonlinear
differential equation,

$$u'(x) = f(x,u(x)), \qquad\qquad (4\text{-}4)$$

$$u(0) = c, \quad 0 \le x \le b.$$

We shall assume f to be continuous in both
x and u and to have bounded second partial derivatives with respect to u for all u and x in
the domain of interest. We also assume f is
strictly convex in u.

The function f can be expanded around an

initial function $u_0(x)$ in a Taylor series,

$$f(x,u_1) = f(x,u_0) + (u_1(x) - u_0(x)) f_u(x,u_0),$$
$$\qquad\qquad\qquad\qquad\qquad\qquad\qquad (4\text{-}5)$$

where the higher order terms have been omitted.
The expression f_u represents the partial derivative of the function f with respect to the
function u "evaluated" at the function $u_0(x)$.

Combining Eq.(4-4) and Eq.(4-5) we obtain
the linear differential equation,

$$u_1'(x) = f(x,u_0(x)) + (u_1(x) - u_0(x)) f_u(x,u_0(x)),$$

$$u_1(0) = c. \qquad\qquad\qquad\qquad (4\text{-}6)$$

Continuing in this way, we are lead to a sequence
of continuous functions $u_n(x)$ determined by the
recurrence relation,

$$u'_{n+1}(x) = (f_u(x,u_n(x)) u_{n+1}(x)$$

$$+ (f_u(x,u_n(x)) - u_n(x) f(x,u_n(x)),$$

$$u_{n+1}(0) = c.$$

(4-7)

It is important to note that the recurrence
relation is in the form of a linear differential equation requiring an initial approximate
solution to begin the iteration process.
Eq.(4-7) encompasses the basic ideas of the
quasilinear technique which we shall now examine.

4.5   EXISTENCE

In this section we will be concerned with
demonstrating that the sequence Eq.(4-7)
does,in fact, exist and further that there is a
common interval (0,b) in which the functions
are uniformly bounded and converge to a solution of Eq.(4-4).

We shall begin by defining a norm over the
space of all functions as,

$$||u_n(x) - c|| = \max_{0 \le x \le b} |u_n(x) - c|.$$

Under the assumption of boundedness of $f(v)$ and
its first two derivatives for finite v, we can
write,

$$\max \left(|f(v)|, |f'(v)|, |1/2)f''(v)| \right) \leq m < \infty,$$

$$|v-c| \leq 1.$$

We are at liberty to select $v_0(x)$ so that,

$$||v_0(x) - c|| \leq 1. \tag{4-8}$$

Once an approximation v_0 is selected, the initial function used to generate the sequence (4-7) is found by integrating both Eq.(4-6) and we have,

$$u_0(x) - c = \int_0^x (f(s,v_0(s))$$

$$+ (u_0(s) - v_0(s))f_u(s,v_0(s)))ds.$$

Therefore

$$||u_0(x)-c|| \leq x(m + m(||v_0(s)-c|| + 1)),$$

and

$$||u_0(x) - c|| \leq \frac{2mb}{(1 - mb)}. \tag{4-9}$$

From Eq.(4-9) we see $||u_0(x) - c|| \leq 1$ provided $0 < x < b < 1/3m$. We now proceed inductively for under the assumption,

$$||u_n(x)-c|| \leq 1 \quad 0 \leq x \leq b.$$

We find, as before,

$$u_{n+1}(x) - c = \int_0^x (f(u_n(x))$$

$$+ (u_{n+1}(x) - u_n(x))f_u(u_n(x))ds,$$

$$= \int_0^x (f(u_n(s)) + ((u_{n+1}(s) - c) - (u_n(s) - c))f_u \, ds.$$

Hence

$$||u_{n+1}(x) - c|| \leq x(m + m(||u_{n+1}(x) - c|| + 1)),$$

$$||u_{n+1}(x) - c|| \leq 2mx/(1-mx),$$

which will not exceed unity provided Eq.(4-9) holds. Thus we see that under the boundedness condition Eq.(4-8), we know a region exists in which the sequence of functions exists and is uniformly bounded.

4.6   CONVERGENCE

A very important property of the sequence is that of convergence. We say a sequence of functions is quadratically convergent if it converges to a limit u(x) and

$$||u_{n+1}(x) - u(x)|| \leq k \, ||u_n(x) - u(x)||^2,$$

where k is independent of n.

To show the condition under which the sequence $u_n(x)$ is quadratically convergent, we consider the quasilinear sequence defined by combining (4-4) and (4-7),

$$(u - u_n)' = f(x,u) - f(x,u_{n-1})$$

$$-(u_n(x) - u_{n-1}(x))f_u(u_n). \tag{4-10}$$

Rewriting (4-10) gives,

$$u'(x)-u_n'(x) = f(u(x)) - f(u_{n-1}(x))$$

$$\tag{4-11}$$

$$+ (u(x) - u_{n-1}(x))f_u(u_{n-1})$$

$$-(u_n(x) - u(x))f_u(u_{n-1}(x)).$$

By the mean value theorem, we can show that

$$f(u) - f(u_{n-1}) -(u_{n-1} - u)f_u(u_{n-1}) ,$$

$$\tag{4-12}$$

$$= (\tfrac{1}{2})(u_{n-1} - u)^2 f_{uu}(v),$$

where $v(x)$ lies between $u_n(x)$ and $u_{n-1}(x)$.
Substituting Eq.(4-12) into Eq.(4-11) and
integrating, gives

$$u_n - u = \int_0^x (\tfrac{1}{2})(u_{n-1} - u)^2 f_{uu}(v)$$

$$+ (u_{n-1} - u)f_u(u_{n-1})).$$

In view of past work, we may write

$$||u_n - u_n|| \le b(m||u_{n-1} - u_{n-1}||^2 + m||u - u_{n-1}||),$$

where $m = \max\limits_{0 \le x \le b} (f_{uu}(v), f_u(v))$.

Therefore we conclude

$$||u_n - u_n|| \le \left(\frac{bm}{1 - bm}\right) ||u_{n-1} - u_{n-1}||^2,$$

which is to say there is quadratic convergence if there is convergence at all. Indeed, if we consider the sequence

$$||u_{n+1} - u_n|| \le k ||u_n - u_{n-1}||^2,$$

$$||u_n - u_{n-1}|| \le k ||u_{n-1} - u_{n-2}||^2,$$

$$\vdots$$

$$||u_2 - u_1|| \le k ||u_1 - u_0||^2.$$

By substitution

$$||u_{n+1} - u_n|| \le k ||u_n - u_{n-1}||^2,$$

$$\le k(k||u_{n-1} - u_{n-2}||^2)^2,$$

$$(4\text{-}13)$$

$$\leq k(k^{2n-2} \; ||u_1 - u_0||^{2n}),$$

$$= (k||u_1 - u_0||)^{2n} /k.$$

If $||u_1 - u_0|| < 1$, then the right side of of Eq.(4-13) approaches zero as n increases and the sequence converges.

We wish to emphasize the importance of quadratic convergence which is associated with the quasilinear technique. Not only does it aid the computing algorithm but it assures rapid convergence to a numerical solution.

4.7   AN EXAMPLE, PARAMETER IDENTIFICATION

Quasilinearization seeks to associate the function f(t) with the "best" approximate differential equation in the interval (0,T). The governing differential equation,in turn, is a mathematical description of the underlying process producing x(t).

A very interesting example of this is the Van der Pol equation

$$x'' + \lambda(x^2 - 1) x' + x = 0, \qquad (4\text{-}14)$$

where the value of the parameter λ clearly influences the behavior of the solution x(t). For example, of $\lambda = 0$,then we have a simple harmonic oscillation.

It is of some interest then to ask whether or not observations of the solution can yield an effective determination of the parameter λ.

Assume that the following three observations of the displacement x(t) have been found.

$$x(4) = - 1.80843,$$

$$x(6) = - 1.63385,$$

$$x(8) = - 1.40456.$$

We wish to determine both λ and x'(4). From the available information we see

$$\frac{x(6) - x(4)}{2} = 0.0.87 = x'(4),$$

$$\frac{x(8) - x(6)}{2} = 0.114 = x'(6),$$

$$\frac{x'(6) - x'(4)}{2} = -.014 = x''(4).$$

Using Eq.(4-14) we get an initial estimate of λ.

$$\lambda = 7.0.$$

The parameter λ in Eq.(4-14) can be considered to be a function of t, subject to the differential equation $\lambda' = 0$. The Van der Pol equation can now be written as a first order system,

$$x' = u ,$$

$$u' = -\lambda (x - 1)^2 - x, \qquad (4-15)$$

$$\lambda' = 0.$$

We can integrate Eq.(4-15) numerically, subject to the approximate initial conditions

$$x(4) = -1.80843,$$

$$x'(4) = 0.087,$$

$$\lambda(4) = 7.0.$$

and compute $x_0(t), u_0(t)\, \lambda_0(t)$ on the interval $4 \leq t \leq 8$.

To obtain the (n+1)st approximation after having calculated the n-th, we use the linearization relation,

$$x'_{n+1} = u_{n+1},$$

$$u'_{n+1} = \lambda_n (x_n^2 - 1)u_n - x_n$$

$$+ (x_{n+1} - x_n)(-2\lambda_n x_n u_n - 1)$$

$$+(u_{n+1} - u_n)(-\lambda_n (x_n^2 - 1))$$

$$+ (\lambda_{n+1} - \lambda_n)(-(x_n^2 - 1)u_n),$$

$$\lambda'_n = 0,$$

(4-16)

where the initial conditions for Eq.(4-16) are estimated as before from $x_n(4), x_n(6)$ and $x_n(8)$.

The results of a numerical experiment are shown in table 4.1.

Table 4.1

Parameter Identification

	init approx	iteration 1	iteration 2	true value
x(4)	-1.80843	-1.80843	-1.80843	-1.80843
u(4)	0.08	0.05644	0.07949	0.07936
λ(4)	7.0	9.91541	10.0004	10.0000

4.8   UNKNOWN INITIAL CONDITIONS

We have yet to exhaust all the opportunities we have in considering the quasilinear equation developed in the last sections. In this section we shall see how the linearity of the equation will allow us to approximate the initial conditions of the differential equation if the solution has been observed over the interval a finite interval.

Let the interval be (t_1, T) where t_1 is close but not equal to the initial point and the differential equation is,

$$u'(t) = f(t, u(t)) . \qquad (4-17)$$

Given a process which has been observed over a time interval (t_1, T) we wish to obtain an approximation to the initial condition $u(0) = c$

from the observed data.

The quasilinear sequence of functions generated by Eq.(4-17) is

$$u'_{n+1} = f(t,u_n(t)) + (u_{n+1}(t)-u_n(t))f_u(t,u_n(t)),$$

or

$$u'_{n+1} = f_u u_{n+1} + f(t,u_n) - f_u u_n .$$

$$(4-18)$$

One sees at once that Eq.(4-18) is a linear differential equation and as such the solution can be written as the sum of a particular and a homogeneous solution, that is

$$u_{n+1}(t) = p_{n+1}(t) + \alpha h_{n+1}(t),$$

$$(4-19)$$

where α is, as yet unknown.

Furthermore both solutions can be computed independently for, we can set the initial conditions, so,

$$p'_{n+1}(t) = f_u(t,u_n) p_{n+1} + f(t,u_n)$$

$$-f_u(t,u_n)u_n ,$$

$$(4-20)$$

$$p_{n+1}(0) = 0,$$

and

$$h'_{n+1}(t) = f_u(t,u_n) h_{n+1} , \qquad (4-21)$$

$$h_{n+1}(0) = 1.$$

Since Eq.(4-20) and Eq.(4-21) are initial value problems with known initial conditions, we can easily integrate (and store) both solutions over the interval (0,T).

We hasten to point out that because of our choice of $p_{n+1}(0) = 0$, we see from Eq.(4-19) that,

$$u_{n+1}(0) = \alpha.$$

The parameter α is unknown, but we now ask the the solution generated by Eq.(4- 19) best fit the known data in the least square sense. Let,

$$J(\alpha) = \int_{t_1}^{T} (u(t) - (p_n(t) + \alpha h_n(t)))^2 \, dt.$$

The parameter α is determined at the nth iteration by the condition $J(\alpha)$ is a minimum.

$$\frac{\partial J(\alpha)}{\partial \alpha} = 0.$$

Calling this value α_n, we now generate

$$u_n(t) = p_n(t) + \alpha_n h_n(t),$$

and we are now in a position to begin the (n+1)st iteration. This continues until an acceptable convergence is achieved.

As an initial guess we can integrate the nonlinear equation Eq.(4-17) using the initial condition extrapolated from the known data.

4.9   DAMPED OSCILLATIONS

A damped oscillatory system is governed by the differential equation,

$$x''(t) + a\, x'(t) + b\, x(t) = 0,$$

$$x(0) = c_1, \qquad (4-22)$$

$$x'(0) = c_2.$$

In this case we will combine the points brought out in the last two sections by asking for both the damping coefficient a, the stiffness coefficient b and the two initial conditions.
We assume the system has been observed over the time interval (t_1, T) and that this information is available. The system can be written as a set of first order differential equations,

$$x' = u,$$

$$u' = -(au + bx),$$

$$a' = 0,$$

$$b' = 0,$$

from which the quasilinear technique yields,

$$x'_{n+1} = u_{n+1},$$

$$u'_{n+1} = -((a_n u_n + b_n x_n) + (a_n (u_{n+1} - u_n)$$

$$+ (u_n (a_{n+1} - a_n)$$

$$+ b_n (x_{n+1} - x_n) + x_n (b_{n+1} - b_n)),$$

$$a'_{n+1} = 0,$$

$$b'_{n+1} = 0.$$

By computing the particular and three homogeneous solutions, we ask for the four constants a_1, b, c and c_2 which minimizes the integral,

$$\int_{t_1}^{T} (y(t) - (p_n(t) + a_1 h_1(t) + a_2 h_1^2(t)$$

$$(4\text{-}23)$$

$$+ a_3 h_1^3(t) + a_4 h_1^4(t)) \, dt,$$

where $y(t)$ is the observed data and the particular and homogeneous solutions are vector solutions having the initial conditions,

$$p_n(0) = \begin{vmatrix} 0 \\ 0 \\ 0 \\ 0 \end{vmatrix}, \; h_1(0) = \begin{vmatrix} 1 \\ 0 \\ 0 \\ 0 \end{vmatrix}, \; h_2(0) = \begin{vmatrix} 0 \\ 1 \\ 0 \\ 0 \end{vmatrix}, h_3(0) = \begin{vmatrix} 0 \\ 0 \\ 1 \\ 0 \end{vmatrix},$$

$$h_4(0) = \begin{vmatrix} 0 \\ 0 \\ 0 \\ 1 \end{vmatrix}$$

where in Eq.(4-23) $h_i^j(t)$ is the jth componant solution of the i-th vector solution. By way of an example, let the known data have the form,

$$y(t) = e^{\beta t}(A \sin \omega t + B \cos \omega t). \qquad (4-24)$$

Substituting Eq.(4-24) into Eq.(4-22) we see that,

$$a = 2\beta,$$

$$b = \beta^2 + \omega^2. \qquad (4-25)$$

In particular let

$$y(t) = e^{-t}(0.07387 \sin 10t + 0.9844 \cos 10t).$$

The results of a numerical study is shown in table 4.2.

Table 4.2

Identification of a System with Damped Oscillations

	init approx	iteration 1	iteration 2
x(0)	.9848	.9652	.9857
x'(0)	-1.7130	-1.1605	-1.7290
a	3.0000	1.6546	2.0224
b	99.0000	103.5824	101.1959

iteration 3	iteration 4	true value
.9848	.9848	.9848
-1.7231	-1.7232	-1.7232
2.0009	1.9999	2.0000
100.9870	100.9995	101.0000

Using Eq.(4-25) it is immediately clear that β = 1 and ω = 10 thereby completing our approximation of the system parameters and the initial conditions.

4.10   SEGMENTAL DIFFERENTIAL APPROXIMATION

Suppose that we observe the output f(t) over
the interval (0,T) from a black box comprising
subsystems,   each subject   to a differential
equation with unknown parameters.It is desired
to approximate the nature of the black box and
if we have control over the input,   this may
then be regarded as an adaptive feedback problem in which the information gathered thus far
gives an indication of what input will now pro-
vide the most additional information.

In this situation we are studying a approximation method which seeks to identify the underlying mathematical structure of the problem
by approximating the output of the system,or
black box.

We now consider the problem of fitting N
different differential approximations in N subintervals of (0,T) where the end points,

$$T_1, T_2 \ldots T_{n-1}, \quad (0 < T_1 < T_2 < \ldots < T_{n-1} = T),$$

are not prescribed so   that determination of
them is part of the optimization procedure.
The set of times then gives estimates of the
times at which the black box switches from one
subsystem to another. This procedure we choose
to call segmental differential approximation
which is a blend of differential approximation
and dynamic programming.

As we have seen before, it is clear that if
a set of data is observed over a time span
and the form of the differential equation is
specified, that the generalized initial conditions and the associated error can be found. It
is reasonable to ask, in the philosophy of dynamic programming for those critical times at
which the differential equation must change in
order to minimize the error over the entire
time span.

Suppose that the entire time interval $(0,T)$ is subdivided into a set of Nk intervals in which a subset defines the true critical times of the system. Further let us assume that within each subinterval we wish to fit differential approximation of constant order M. Thus, in the kth subinterval, we have the function given by,

$$u_k^M + b_{1k} u_k^{M-1} + \ldots + b_{Mk} u_k = 0,$$

where $u_k^M = \dfrac{d^M}{dt^M} u_k(t)$.

in $t_k \leq t \leq t_{k+1}$ we have the initial conditions

$$u_k^j(t_k) = c_{jk} , \quad j = 0,1,\ldots M-1.$$

Then if

$$S_A(b_{ij}, c_{ij}, t_k) = \int_{t_k}^{t_{k+1}} (u_k(t) - f_k(t))^2 \, dt,$$

where $f_k(t)$ is the observed output of the black box over the interval $t_k \leq t \leq t_{k+1}$. We wish to determine b_{jk}, c_{jk}, and t_k that gives

$$F_N(T) = \min_{(b_{jk}, c_{jk}, t_k)} S_A(b_{jk}, c_{jk}, t_k),$$

with

$$F_1(T) = S_A(0,0,T),$$

where $S_A(0,0,T)$ is the sum of the squared differences for the optimal differential approximation in the N th subinterval.

4.11   DIFFERENTIAL SYSTEMS WITH TIME VARYING COEFFICIENTS

Let $f(t)$ be observed over the interval $(0,T)$. Suppose it is defined by the first order linear differential equation,

$$f'(t) = a(t) f(t) + b,$$

where the coefficient $a(t)$ is itself subject to a feedback equation,

$$a(t)= b_0 + b_1 f(t) + b_2 f'(t) + \ldots + b_{(k+1)} f^{(k)}(t),$$

where $f^{(k)}$ is the k-th derivative of f.

Alternatively, in the interval $(0,T_0)$, we wish to determine a function $u(t)$ defined by constants $b_0, b_1, \ldots b_k$ and the initial conditions,

$$u^{(j)}(0) = c_j, \quad j = 0,1,\ldots,k-1,$$

and by the differential equation

$$u'(t) = a(t) \, u(t) + b, \qquad (4\text{-}26)$$

where

$$a(t) = b_0 + b_1 \, u(t) + b_2 \, u'(t) + \ldots$$
$$+ b_{(k+1)} \, u^{(k)}(t), \qquad (4\text{-}27)$$

that minimizes,

$$S = \int_0^T (u(t) - g(t))^2 \, dt,$$

where $g(t)$ is observed over the interval $(0,T)$. That is, b_j and c_j are chosen to minimize the square difference between the theoretical function $u(t)$ and the observation $g(t)$.

4.12   A METHOD OF   SOLUTION

Using Eq.(4-26)   and   Eq.(4-27),   we obtain the equation

$$b_{k+1} \, u^{(k)} = \frac{u' - b}{u} - b_0 - b_1 u' - \ldots$$
$$- b_k \, u^{(k-1)}. \qquad (4\text{-}28)$$

Continuity of the first k derivatives of u(t) will be assumed so that,   for example, when u =0,   the first term of the right side of Eq.(4-28) is considered to be,

$$\lim_{u \to 0} \frac{u' - b}{u} .$$

If we further assume $b_0, b_1, ..b_{k+1}$ to be the state variables we arrive at the system of differential equations.

$$\dot{u}^{(0)} = u^{(1)},$$

$$\dot{u}^{(1)} = u^{(2)},$$

$$\vdots$$

$$\dot{u}^{(k-2)} = u^{(k-1)},$$

$$\dot{u}^{(k-1)} = \frac{1}{b_{k+1}} (\frac{u^{(1)} - b}{u} - b_0 - b_1 u^{(2)}$$

$$... - b_k u^{(k-1)}),$$

$$\dot{b} = 0,$$

$$\vdots$$

$$\dot{b}_{k+1} = 0.$$

Now , if we put

$$u^{(0)} = x_0,$$

$$u^{(1)} = x_1,$$

$$\vdots$$

$$u^{(k-1)} = x_{(k-1)},$$

$$b = x_k,$$

$$b0 = x_{(k+1)},$$

$$\vdots$$

$$b^{.(k+1)} = x_{(2k+2)},$$

then the nonlinear system of differential equations in the variable $x = (x_0, x_1, \ldots x_{(2k+2)})$ is given by,

$$\dot{x} = g(x).$$

As before, we now quasilinearize, obtaining successive solutions , $x_1(t)$, $x_2(t)$ ... .
The set of linear differential equations at the nth stage are,

$$\dot{x}_j^{(n)} = g_j(x^{(n-1)})$$

$$+ \sum_{m=0}^{2k+2} \frac{\partial g_j(x^{(n-1)})}{\partial x_m}(x_m^{(n)} - x_m^{(n-1)}).$$

Following the previous procedure, we determine (2k+2) independent homogeneous vector solutions and one particular solution. The n-th quasilinear vector solutions is then the linear combination

$$u^{(n)} = p(t) + \sum_{j=0}^{2k+2} \alpha_j h_j(t).$$

The coefficients α_j are determined by the requirement that,

$$\int_0^T (f(t) - (p(t) + \sum_{j=0}^{2k+2} \alpha_j h_j(t)))^2 dt,$$

is a minimum, exactly as before.

4.13   AN INTERESTING CASE

Suppose that instead of Eq.(4-26) and Eq.(4-27) we have,

$$u'(t) = a(t) u(t) + b , \qquad (4-29)$$

as before,but now we change the feedback condition,

$$\dot{a}(t) = ca + b_0 + b_1 u + \dots b_{(k+1)} u^{(k)} .$$

Then $u^{(k)}$ can be expressed as   (4-30)

$$u^{(k)} = \frac{1}{b_{k+1}} \left(\frac{u^{(2)} - au^{(1)}}{} - ca - b_0 - \ldots - b_1 u^{(k-1)} \right),$$

by differentiating Eq.(4-29) and using Eq.(4-30).

$$a = \frac{u^{(1)} - b}{u},$$

so from Eq.(4-29)

$$u^k = \frac{1}{b_{(k+1)}} \left(\frac{u^{(2)} u^{(1)} - u^{(1)}(u^{(1)} - b)}{u^2} - \frac{c(u^{(1)} - b)}{u} \right.$$

$$\left. -b_0 - b_1 u - \ldots - b_k u^{(k-1)} \right).$$

Hence, putting

$$u^{(0)} = x_0,$$

$$\vdots$$

$$u^{(k-1)} = x_{(k-1)},$$

$$b_k = x_k,$$

$$b0 = x_{(k+1)},$$

$$\vdots$$

$$b_{k+1} = x_{(2k+2)},$$

$$c = x_{(2k+3)}.$$

we again obtain a nonlinear differential equation,

$$\dot{x} = g(x)$$

for the state vector $x = (x_0, x_1 \ldots x_{(2k+3)})$ and proceed exactly as before.

4.14   DISCUSSION

Modern methods of approximations need not be restricted to the conventional representation of data. By quasilinearization we have seen that we can also obtain an approximation of the underlying structure of the basic process by determining the form of the governing differential equations.

Like all approximation methods, we do not present the differential approximation technique without some words of caution. Having the observed data only, it can be very difficult to chose the exact form of the differential equation. However, since the method yields an associated absolute error, one could roam over several sets of differential forms to find the form giving the lowest error.

The choice of the initial approximation is
very important and care must be exercised in
its selection. In approximating the initial
conditions several points must be made. In propagation problems the effect of the initial
conditions on the solution may damp out quickly
during the transient portion and disappear in
the steady state solution. Therefore in this
situation, we must observe the system over a
shorter period of time but with extreme accuracy. Finally, a point of interest. The observed
data, which drives the approximation technique,
need not be observed data, although this is a
reasonable assumption. We could, just as well,
use design data thereby using a mathematical
approximating method as an engineering design
tool.

4.15   BIBLIOGRAPHY AND COMMENTS

In 4.1 , for the basic ideas of quasilinearization, see,

> Bellman,R.:1970, Methods of Nonlinear
> Analysis,vol I & II ,Academic Press,
> N.Y.

> Bellman,R. and R.Kalaba:1965, Quasi-
> linearization and Nonlinear Boundary
> Value Problems, American Elsevier
> Publishing Co,. N.Y.

In 4.6 for further analysis of convergence see,

> Kalaba,R.:1954,"On Nonlinear Different-
> ial Equations,the Maximum Operation and
> Monotone Convergence"
> J. Math. Mech,8,519-574

In 4.9, we use the result,

> Roth,R.:1966,"Data Unscrambling: Studies

in Segmental Differential Approximation',
JMAA,14,1,5-22

Additional reading can be found in.

Bellman,B. Gluss and R.S.Roth,:1964,"On
the Identification of Systems and the
Uscrambling of Data: Some problems
suggested by Neurophysiology"'
Proc.Nat.Acad.Sci,52,1239-1240

Bellman,R.,B.Gluss and R.S.Roth:1965,"
"Segmental Differential Approximation
and the 'Black Box' Problem" JMAA,12,
191-204

Bellman,R. and R.S.Roth,:1966,"A Technique for the Analysis of a Broad Class
of Biological Systems",   Bionics Symp,
Gordon and Breach

Bellman R, and R.S.Roth,:1966,
"Segmental Differential Approximation
and Biological Systems:An Analysis of
a Metabolic Process",J. Theor. Biol.
11,168-176

Roth,R.S. and M.M,Roth,:1969,"Data
Unscrambling and the Analysis of
Inducible Enzyme Synthesis", Math.
Biosci,5,57-92.

Chapter 5

DIFFERENTIAL APPROXIMATION

5.1 INTRODUCTION

In the last chapter we considered the technique of quasilinearization for fitting a known function $f(x)$ to a differential equation by determining the initial conditions and system parameters associated with the chosen differential equation. This was done by numerical methods.

In this chapter we wish to explore the analytical methods for doing a slightly different approximation. In this case we are given a nonlinear differential equation and we seek to determine an associated linear differential equation whose exact solution is a good approximation to the solution of the original equation in the range of interest.

5.2 DIFFERENTIAL APPROXIMATION

Let $T(u,b)$ be a class of operators dependent on a vector parameter, b. Our aim is to determine b so as to minimize the functional,

$$|| T(v,b) ||,$$

where $||.||$ is a suitably chosen norm. We expect to find that a suitably chosen solution of

$$T(v,b) = 0,$$

will furnish a approximation to u.

5.3   LINEAR DIFFERENTIAL OPERATORS

One of the most important classes is the linear differential operator with constant coefficients.

Let,

$$T(u,b) = u^{(n)} + b_1 u^{(n-1)} + \ldots + b_n u.$$

Let us consider the problem of minimizing the expression,

$$\int_0^T (u^{(n)} + b_1 u^{(n-1)} + \ldots + b_n u)^2 \, dt,$$

with respect to the b's where $u(t)$ is a known function. If we assume that the functions $u, u', u'', \ldots$ are linearly independent, the minimizating values are uniquely determined as the solution of a system of algebraic equations,

$$\int_0^T (u^{(n)} + b_1 u^{(n-1)} + \ldots b_n u) u^{(n-i)} \, dt = 0,$$

$$\quad (5-1)$$

$$i = 1, 2, \ldots n.$$

We make a note of cautious optimism here by observing that as long as $u_n, u_{n-1} \ldots u_1$ are integrable and linearly independent the set of Eq.(5-1) will yield b_i to best approximate $u(t)$ in the least square sense. Only by solving the equation,

$$T(v,b) = 0,$$

b_i, $i=1,2,\ldots n$, being found and selecting the proper end (or initial) conditions can the approximation be verified.

5.4 DEGREE OF APPROXIMATION

Consider the expression obtained in the last section.

$$\int_0^T (u^{(n)} + b_1 u^{(n-1)} + \ldots + b_n u)^2 \, dt \quad (5\text{-}2)$$

where u is a known function and b_i are determined by minimizing Eq.(5-2). Associated with the functional is the linear differential equation,

$$v^{(n)} + b_1 v^{(n-1)} + \ldots + b_n v = 0,$$

where we set $u^{(k)}(0) = v^{(k)}(0)$, $k = 0,1,\ldots$ n-1. The function u satisfies the equation,

$$u^{(n)} + b_1 u^{(n-1)} + \ldots + b_n u = h(t),$$

where b_i have been determined so that $||h(t)|| < \epsilon$.
We would hope that by this process v(t) would be close to u(t).

As n gets large it is very difficult to construct a general argument showing that v becomes an arbitrarily good approximation to u. Indeed, as we shall see, the differential approximation technique is meaningful only if n is of moderate size.

5.5   IMPROVING THE APPROXIMATION

Recalling that the differential approximation method seeks to find a differential equation whose solution is a good approximation to
the known function, we are led to a function v
which is a solution to,

$$v^{(n)} + b_1 v^{(n-1)} + \ldots + b_n v = 0, \qquad (5\text{-}3)$$

$$v^{(k)}(0) = u^{(k)}(0),$$

$$k = 0,1,2\ldots n-1.$$

Now we ask if there is another solution of
Eq.(5-3) which yields a better approximation in
the sense of decreasing the quantity,

$$||v - u||^2 = \int_0^T (u - v)^2 \, dt . \qquad (5\text{-}4)$$

This can be done by loosening the restrictions
on the initial conditions. Let $v_1, v_2, \ldots v_n$
be the characteristic solutions of Eq.(5-3) .
Now we can set,

$$v = \sum_{i=1}^{n} a_i v_i,$$

where the a's are chosen to minimize Eq.(5-4).
We could also construct the set of mutually orthogonal solutions of Eq.(5-3) , which makes
the determination of the a's very easy.

5.6   AN EXAMPLE

Consider the equation,

$$u(t) = (1 - \int_0^t e^{-s^2} ds) + \int_0^t e^{-(t-s)^2} u(s) ds$$

$$(5-5)$$

having the solution $u(t) = 1$.
Let the kernel of Eq.(5-5) be,

$$k(t) = e^{-t^2},$$

and the forcing function be,

$$f(t) = 1 - \int_0^t e^{-s^2} ds.$$

Equation Eq.(5-5) can be written as,

$$u(t) = f(t) + \int_0^t k(t-s)u(s) ds. \qquad (5-6)$$

We wish to find an approximate solution to Eq.(5-6). Let us begin by differentiating Eq.(5-6) three times so that

$$u'(t) = f'(t) + k(0)u(t) + \int_0^t k'(t-s)u(s)ds,$$

$$(5-7)$$

$$u''(t) = f''(t) + k(0)u'(t) + k'(0)u(t)$$

$$(5-8)$$

$$+ \int_0^t k''(t-s)u(s)ds,$$

$$u'''(t) = f'''(t) + k(0)u''(t) + k'(0)u'(t)$$

$$+ k''(0)u(t)$$

$$+ \int_0^t k'''(t-s)u(s)ds. \qquad (5-9)$$

Let us take a linear combination of the terms defined above,

$$u''' + b_2 u'' + b_1 u' + b_0 u =$$

$$f''' + b_2 f'' + b_1 f' + b_0 f + b_1 k(0)u(t) \qquad (5-10)$$

$$+ b_2 (k(0) u'(t) + k'(0) u(t))$$

$$+ (k(0)u''(t)+k'(0)u'(t)+k''(0)u(t))$$

$$+ \int_0^t (k'''(t-s) +b_2 k''(t-s) + b_1 k'(t-s)$$

$$+ b_0 k(t-s))u(s)ds.$$

We now observe that if f satisfies a linear differential equation with constant coefficients, u will satisfy an inhomogeneous differential equation where the initial condition can be found from Eqs.(5-7 - 5-9).

But k(t) is the known kernel of Eq.(5-5). Therefore, we can approximate k by the linear differential equation.

$$k''' + b_2 k'' + b_1 k' + b_0 k = 0,$$

where the b's are determined by the condition,

$$\int_0^1 (k''' + b_2 k'' + b_1 k' + b_0 k)^2 \, dt,$$

is a minimum.

we can determine the value of b_1, b_2, b_3, which are

$$b_1 = 2.7402990,$$
$$b_2 = 7.9511452,$$
$$b_3 = 5.7636455.$$

The function

$$f(t) = 1 - \int_0^t e^{-s^2} \, ds,$$

satisfies the third order differential linear equation,

$$f^{(3)} + 2t\, f^{(2)} + 2\, f^{(1)} = 0,$$

$$f(0) = 1, \quad (5\text{-}11)$$
$$f'(0) = -1,$$
$$f''(0) = 0.$$

Eq. (5-5) is replaced by two simultaneous differential equations, one for u, obtained by

ignoring the integral in Eq.(5-10) and the
other, an equation for f, given by Eq.(5-11).

Since the coefficients and the initial conditions are known, the ideas of differential
approximation allows us to integrate the two
equations for f and u over the interval
(0,1), u(t) is then the approximate solution to
Eq.(5-5).

Finally we note that by fitting the system
to higher derivatives, better accuracy can be
obtained.

5.7   DIFFERENTIAL-DIFFERENCE EQUATIONS

Now let us consider the approximate solution
of the differential- difference equation,

$$u'(t) = g(u(t-1),u(t)), \quad t > 1,$$

$$u(t) = h(t) \quad 0 < t < 1. \tag{5-12}$$

Let us introduce a sequence of functions,

$$u_n(t) = u(t + n). \tag{5-13}$$

The Eq.(5-12) can be written as an infinite
system of ordinary differential equations.

$$u_n'(x) = g(u_n, u_{n-1}), \tag{5-14}$$

$$0 < t < 1$$

$$n = 0,1 \ldots ,$$

where we must determine the initial conditions
for each equation. We know immediately,

$$u_0(0) = u(0) = h(0),$$

$$u_1(0) = u(1) = h(1).$$

Now we can easily integrate the equation,

$$u_1'(t) = g(u_1, u_0),$$

$$= g(u_1, h(t)),$$

numerically, to obtain $u_1(1) = u_2(0)$. After having determined the new initial condition, we can integrate two equations,

$$u_1'(t) = g(u_1, h(t)), \quad u_1(0) = h(1),$$

$$u_2'(t) = g(u_2, u_1), \quad u_2(0) = u_1(1).$$

To obtain $u_2(1) = u_3(0)$, we now integrate the set of three equations,

$$u_1'(t) = g(u_1, h(t)), \quad u_1(0) = h(1),$$

$$u_2'(t) = g(u_2, u_1), \quad u_2(0) = u_1(1),$$

$$u_3'(t) = g(u_3, u_2), \quad u_3(0) = u_2(1),$$

and so on. This method, however, leads to an embarrassing number of equations. To avoid this, we use the ideas of differential approxi-

mation.   If   we   use   the   foregoing
technique,together with Eq.(5-12),   we can obtain an approximate function which   satisfies
the differential equation,

$$\frac{dv}{dt} = F(v) \ , \ 0 < t < 1,$$

where v depends, perhaps implicitly, on n.   We
then start over again integrating,

$$u'(t) = g(u(t),u(t-1)), \ t> n+1,$$

$$u(t) = v(t), \qquad n < t < n+1.$$

By converting a   single differential-difference equation into an   ever increasing sequence
of initial value problems,   we have avoided the
problem of having to store the   initial condition h(t),very accurately, in fast storage over
the interval (see Eq. 5-12). The advantage of
saving fast storage,which   will become increasingly important when small personal computers
are used to solve large problems,   is offset by
larger computing time in solving a very system
of differential equations.

5.8   A USEFUL APPROXIMATION TO G(T)

Let

$$g(t) = e^{-t^2}.$$

The function g(t) occurs in many important
contexts in mathematics.In some it is quite
useful to replace it by an approximation of
some type,   such as a Pade approximation. In
this section we wish to exhibit a surprisingly

good approximation as the sum of three exponentials. This is obtained by using differential approximation which hold for $0 < t < 1$.

Given a function $v(t)$ for $0 < t < T$, we determine the coefficients a_i which minimizes the quadratic expression,

$$J(a_i) = \int_0^T (v^{(n)} + \sum_{i=1}^{n} a_i v^{(n-i)})^2 \, dt, \qquad (5\text{-}15)$$

where $v^{(k)}$ is the k-th derivative of v.
We the expect that the solution of the linear differential equation,

$$u^{(n)} + a_1 u^{(n-1)} + \dots + a_n u = 0, \quad (5\text{-}16)$$

with suitable boundary conditions will yield an approximation to $v(t)$. If we let,

$$v(t) = e^{-t^2}, \qquad 0 < t < 1,$$

then Eq.(5-15) for $n = 3$, becomes,

$$J(a_i) = \int_0^T (v^{(3)} + a_1 v^{(2)} + a_2 v^{(1)} + a_3 v)^2 \, dt.$$

The minimizing set a_i then are the coefficients of the linear differential equation,

$$u^{(3)} + a_1 u^{(2)} + a_2 u^{(1)} + a_3 u = 0. \qquad (5\text{-}17)$$

As initial conditions, select,

$$u^{(2)}(0) = v^{(2)}(0).$$

If we express the solution of the linear differential equation Eq.(5-17) as the sum of exponentials, we obtain,

$$u(t) = \sum_{i=1}^{n} b_i e^{-\lambda_i},$$

where b_i and λ_i may be complex. The values are calculated and listed in table 5.1 The numerical values of the function u(t) are given in table 5.2

Table 5.1

Calculated Parametric Values

a_i	b_i	λ_i
2.7403	0.7853	0.9180
7.9511	0.1074 + .1963i	0.9111 + 2.334i
5.7636	0.1074 - .1963i	0.9111 - 2.2334i

Table 5.2

Results of an Approximation		
time	calculated value	absolute error
0.1	0.99003	$0.300 \ 10^{-4}$
0.3	0.913676	$0.255 \ 10^{-3}$
0.5	0.778679	$0.122 \ 10^{-3}$
0.8	0.527665	$0.372 \ 10^{-3}$
1.0	0.367951	$0.724 \ 10^{-4}$

5.9   DISCUSSION

If desired, we can improve the accuracy of the approximation by taking the values of $u_i(0)$, the initial conditions, as parameters, and determining these values by minimizing the quadratic,

$$J(c_i) = \int_0^T \left(v(t) - \sum_{i=1}^{n} c_i u_i \right)^2 dt,$$

$$u^{(i)}(0) = c_i,$$

where $u_1, u_2, \ldots u_n$ are n linearly independent solutions of Eq.(5-16).

The integrals which arise are evaluated by using the differential equation Eq.(5-16) and the auxiliary conditions,

$$\frac{d h_{ij}}{dt} = u_i u_j, \quad h_{ij}(0) = 0,$$

$$\frac{d w_j}{dt} = u_j v, \quad w_j(0) = 0,$$

then

$$h_{ij}(T) = \int_0^T u_i u_j \, dt, \quad w_j = \int_0^T u_j v \, dt.$$

The same techniques can often be used in the determination of the coefficients where the function $v(t)$ satisfies a differential equation, linear or nonlinear. In this case

$$v' = -t^2 v \quad, \quad v(0) = 1.$$

5.10   AN EXAMPLE

Consider the equation

$$u'(t) = -u(t-1)(1 + u(t)), \quad t > 1$$

$$u(t) = 1. \qquad\qquad 0 < t < 1.$$

By the technique discussed in section 5.7, we have a system of ordinary differential equations,

$$u_0'(t) = 0, \qquad\qquad u_0(0) = 1$$

$$u_1'(t) = -u_0(1 + u_1), \quad u_1(0) = u_0(1),$$

$$\vdots$$

$$u_k'(t) = -u_{k-1}(1 + u_k), \quad u_k(0) = u_{k-1}(1).$$

$$(i)$$

Let $z(t) = u_n(t)$, $z_i - z$, $z_0 = z$, then

$$z_0 = u_n,$$

$$z_1 = u_n',$$

$$z_2 = u'_{n-1}(1 + u_n) - u_{n-1}u'_n = u''_n,$$

$$z_3 = u''_{n-1}(1 + u_n) - 2u'_{n-1}u'_n - u_{n-1}u''_n = u'''_n,$$

where $u''_{n-1} = -u'_{n-2}(1 + u_{n-1}) - u_{n-2}u'_{n-1}$.

In the following numerical calculation, a grid size of 0.00390625 was chosen an integration begun at t= 1 and ended at t = 20. After t = 4k mins.,k = 1,2,3,4 the differential approximation of order R = 1,2,3 was studied.

Table 5.3 compares u(t) using differential approximation with its exact values for t = 4,6,....18.

Table 5.3

	Approximating the Solution of		
	a Differential-Difference Equation		
t	u(t) exact	u(t) R = 1	u(t) R = 2	U(t) R = 3
4	0.16534	0.16838	0.16533	0.16534
6	-0.05369	-0.05267	-0.05369	-0.05369
8	-0.00455	-0.00606	-0.00454	-0.00455
10	0.01657	0.01762	0.01657	0.01657
12	-0.01477	-0.01533	-0.01477	-0.01477
14	0.00932	0.00956	0.00932	0.00932
16	-0.00466	-0.00473	-0.00466	-0.00466
18	0.00179	0.00179	0.00179	0.00179

5.11   FUNCTIONAL DIFFERENTIAL EQUATIONS

The ideas of differential approximation can be applied to functional differential equations of the form,

$$u'(t) = g(u(t), u(h(t))). \qquad (5-18)$$

Let $h(t) < t$ for $t > 0$, so that the future is determined solely by the past. Further suppose $h'(t) > 0$ for $t > 0$ and let $H(t)$ be the inverse function of $h(t)$. Take $u(t)$ to be known in some initial interval $(0, t)$.

Let the sequence t_n be, where $t_1 = H(0)$,

$$t_n = H(t_{n-1}), \quad n = 2, 3 \ldots$$

Then $t_n = H^{(n)}(0)$, the n-th iterate of $H(t)$ evaluated at $t=0$. We also see that $H(t)$ maps the interval (t_{n-1}, t_n) onto (t_n, t_{n+1}) in a one-to-one fashion, and the k-th iterate maps (t_{n-1}, t_n) onto (t_{n-1+k}, t_{n+k}).

Consider the sequence of functions,

$$u_n(s) = u(H^{(n)}(s)), \quad n = 0, 1, 2$$

where $0 < s < t <$ and $H(s) = s$.
Then by virtue of Eq.(5-18),

$$u'_n(s) = (\frac{d}{ds} H^{(n)}(s)) u'(H^{(n)}(s)).$$

where the derivative of $H^{(n)}(s)$ may be evaluated by the expression,

$$\frac{d}{ds} H^{(n)}(s) = H'(H^{(n-1)}(s)) \frac{d}{ds} H^{(n-1)}(s).$$

$$(5-19)$$

Now set $t = H^{(n)}(s)$ with $0 < s < t_1$, $n = 1, 2 \ldots$ then Eq.(5-18) yields

$$u'(H^{(n)}(s)) = g(u(H^{(n)}(s)), u(H^{(n-1)}(s))) \quad (5-20)$$

$$= g(u_n(s), u_{n-1}(s)).$$

Hence from Eq.(5-19) and Eq.(5-20) we have,

$$u'_n(s) = (\frac{dH^{(n)}(s)}{ds}) g(u_n(s), u_{n-1}(s)),$$

$n=1,2\ldots$. We proceed to obtain a numerical solution of this equation.

5.12   THE NONLINEAR SPRING

Consider the equation for the nonlinear spring,

$$u'' + u + \epsilon u^3 = 0,$$

$$u(0) = 1,$$

$$u'(0) = 0.$$

We wish to determine an approximate linear equation,

$$v'' + (1 + \epsilon b) v = 0,$$

$$v(0) = 1,$$

$$v'(0) = 0,$$

where b is constant and to compare u and v in the interval $(0,\pi)$. We can obtain a value of b accurate to $O(\epsilon)$ by minimizing,

$$\int_0^\pi (u + \epsilon u^3 - (u + \epsilon bu)^2)\, dt\ ,$$

(5-21)

$$= \epsilon^2 \int_0^\pi (u^3 - bu)^2\, dt.$$

If ϵ is small, we can approximate u by u = cos t . If u is substituted in Eq.(5-21), the value of b which minimizes this expression can be shown to be , b= 3/4.

The approximating equation is then

$$v'' + (1 + \frac{3\epsilon}{4})v = 0,$$

$$v(0) = 1,$$

$$v'(0) = 0,$$

whose exact solution is

$$v = \cos((1 + \frac{3\epsilon}{4})^{1/2} t).$$

5.13   THE VAN DER POL EQUATION

Let us finally consider the Van der Pol equation,

$$u'' + \epsilon(u^2 - 1)u' + u = 0,$$

and see if we can obtain an approximation to the unique periodic solution over $(0,2\pi)$ for small ϵ.
For the approximation, we choose,

$$(u^2 - 1)u' = b_1 u' + b_2 u,$$

and determine b_1 and b_2 by minimizing,

$$\int_0^{2\pi} ((u^2 - 1)u' - b_1 u' - b_2 u)^2 \, dt . \quad (5\text{-}22)$$

The variational equations are,

$$\int_0^{2\pi} ((u^2 - 1)u' - b_1 u' - b_2 u)u' \, dt = 0,$$

$$(5\text{-}23)$$

$$\int_0^{2\pi} ((u^2 - 1)u' - b_1 u' - b_2 u) u \, dt = 0.$$

$$(5\text{-}24)$$

Eq.(5-24) reduces to

$$0 = \int_0^{2\pi} ((u^2 - 1) u'u - b_1 uu') \, dt,$$

$$= b_2 \int_0^{2\pi} u^2 \, dt,$$

so

$$b_2 = 0.$$

Eq.(5-23) then yields,

$$\int_0^{2\pi} (u^2 - 1) u'^2 \, dt = b_1 \int_0^{2\pi} u'^2 \, dt. \qquad (5\text{-}25)$$

For ϵ small, we let $u = k \cos t$ where k is to be determined. Since the approximating equation is,

$$v'' + b_1 v' + v = 0, \qquad (5\text{-}26)$$

it is reasonable to determine k when $b_1 = 0$, or,

$$\int_0^{2\pi} (k^2 \cos^2 t - 1) \sin^2 t \, dt = 0 , \qquad (5\text{-}27)$$

Solving Eq.(5-27) for k yields $k = 4$.

We are now in a position to find the unknown constant from Eq.(5-25) and finally the approximation to the Van der Pol equation is found by solving Eq.(5-26).

5.14   BIBLIOGRAPHY AND COMMENTS

For a general discussion of differential approximation see,

Bellman,R.,:1970, Methods in Nonlinear Analysis,vol I & II, Academic Press, N.Y.

Bellman,R., and K.L.Cooke,:1963, Differential-Difference Equations , Academic Press, N.Y.

Bellman,R.,R.Kalaba and R.Sridhar:1965, "Adaptive Control via Quasilinearization and Differential Approximation",Pakistan Engineer,5,2,94-100

Bellman,R.,B.G.Kashef and R. Vasudevan, :1972, "Application of Differential Approximation in the Solution of Integro-Differential Equations', Utilita Mathematica, 2,283-390

Bellman,R.,B.G.Kashef and R.Vasudevan, :1972, "A Useful Approximation to
e^{-t^2} ",Mathematics of Computation,26,117, 233-235

Chapter 6

DIFFERENTIAL QUADRATURE

6.1 INTRODUCTION

In this chapter we wish to explore yet another application of approximation methods, the numerical solution of nonlinear partial differential equations. In many cases all that is desired is a moderately accurate solution at a few points which can be computed rapidly.

Standard finite difference techniques have the characteristic that the solution must be calculated at a large number of points in order to obtain moderately accurate results at the points of interest. As we will show, differential quadrature is an approximation technique which could prove to be an attractive alternative in the solution of partial differential equations.

6.2 DIFFERENTIAL QUADRATURE

Consider the nonlinear partial differential equation,

$$u_t(x,t) = g(x,t,u,u_x(x,t)), \qquad (6-1)$$
$$-\infty < x < \infty , \ t > 0,$$

with the initial condition,

$$u(x,0) = h(x) . \qquad (6-2)$$

Let us assume that the function $u(x,t)$ satisfying Eq.(6-1) and Eq.(6-2) is sufficiently smooth to allow us to write,

$$u_x(x_i,t) = \sum_{j=1}^{N} a_{ij}\, u(x_j,t), \qquad (6-3)$$

$$i = 1,2 \ldots, N,$$

where the coefficients will be determinded later. Substituting Eq.(6-3) into Eq.(6-1), yields a set of N ordinary coupled nonlinear differential equations,

$$u_t(x_i,t) = g(x_i,t,u(x_i,t), \sum_{j=1}^{N} a_{ij} u(x_j,t)),$$

with initial conditions,

$$u(x_i,0) = h(x_i),$$

$$i = 1,2,\ldots n.$$

Hence, under the assumption that Eq.(6-3) is valid, we have succeeded in reducing the task of solving Eq.(6-1) to that of solving a set of N ordinary differential equations with prescribed boundary conditions.

6.3   DETERMINATION OF THE WEIGHTING
       COEFFICIENTS

In order to determine the coefficients in the approximation

$$f'(x_i) = \sum_{j=1}^{N} a_{ij} f(x_j) \qquad (6-4)$$

$$i = 1,2 \ldots n,$$

we proceed by analogy with the quadrature case
and ask that Eq.(6-4) be exact for all   polynomials of degree less than or equal to   N-1
test functions f(x) = x ,   k = 1, ... N-1,
                        k
leads to a set of linear algebraic   equations,

$$\sum_{j=1}^{N} a_{ij} x_j^{k-1} = (k-1) x_i^{k-2}, \qquad (6-5)$$

$$i = 1,...,N,$$

$$k = 1,...N,$$

which has a unique solution since the coefficient matrix is a Vandermonde matrix.

Instead of inverting a large matrix, which
is required in the solution of Eq.(6-5), we
will choose a different test function.

Let,

$$f(x) = \frac{P_n^*(x)}{(x - x_k) P_N^{*\prime}(x_k)},$$

where $P_N^*(x)$ is defined in terms of the Legendre
polynomials by the relation,

$$P_N^*(x) = P_N(1 - 2x),$$

where $P_N(x)$ is the Nth order Legendre polynomial for $-1 \leq x \leq 1$. By choosing x_i to be the roots of the shifted Legende polynomial we see that

$$a_{ik} = \frac{P_N'^*(x_i)}{(x_i - x_k) P_N'^*(x_k)} \quad , \quad i \neq k.$$

For the case when $i = k$, use of L'Hospital's rule plus the fact that the Legendre polynomial satisfies the differential equation,

$$x(1 - x^2)P_N''^*(x) + (1 + 2x)P_N'^*(x)$$
$$+ N(N+1)P_N^*(x) = 0,$$

gives,

$$a_{kk} = \frac{(1 - 2x_k)}{2x_k(x_k - 1)} .$$

Therefore, by choosing N, the order of the approximation, the N roots of the shifted Legendre polynomial are known. This ,in turn, defines the required coefficients which make the differential quadrature possible.

6.4   A FIRST ORDER PROBLEM

In this section, we shall consider a problem arising in the theory of Dynamic Programming.

Let us consider the nonlinear partial differential equation,

$$u_t(x,t) = x^2 - (u_x^2(x,t))/4, \quad (6-6)$$

$$u(x,0) = 0.$$

The analytical solution to Eq.(6-6) is,

$$u(x,t) = x^2 \tanh(t).$$

Replacing the derivative term in the right side of Eq.(6-6) by an approximating sum, we obtain the set of nonlinear ordinary differential equations,

$$u_t(x_i,t) = x_i^2 - (\frac{1}{4}) (\sum_{j=1}^{N} a_{ij} u(x_j,t))^2.$$

$$(6-7)$$

The system Eq.(6-7) was integrated from t=0 to t = 1 using an Adams-Moulton integration scheme with a step size of Δ = 0.01. The order of the quadrature was chosen to be, N = 7. The results are shown in table 6.1.

Table 6.1

Approximating the Solution of a
Non-linear Partial Differential Equation

---

t	x	computed solution	actual solution
	x_1	$6.4535127 \ 10^{-5}$	$6.4535123 \ 10^{-5}$
0.1	x_4	$2.4916991 \ 10^{-2}$	$2.416993 \ 10^{-2}$
	x_7	$9.4660186 \ 10^{-2}$	$9.4660196 \ 10^{-2}$
	x_1	$2.9922271 \ 10^{-4}$	$2.9922131 \ 10^{-4}$
0.5	x_4	$1.1552925 \ 10^{-1}$	$1.1552926 \ 10^{-1}$
	x_7	$4.3889794 \ 10^{-1}$	$4.3889818 \ 10^{-1}$
	x_1	$4.9314854 \ 10^{-4}$	$4.9313300 \ 10^{-4}$
1.0	x_4	$1.9039847 \ 10^{-1}$	$1.9039851 \ 10^{-1}$
	x_7	$7.2332754 \ 10^{-1}$	$7.2332808 \ 10^{-1}$

---

where for N = 7,

$$x_1 = 2.5446043 \ 10^{-2}$$

$$x_2 = 1.2923441 \ 10^{-1}$$

$$x_3 = 2.9707742 \ 10^{-1}$$

$$x_4 = 5.0000000 \ 10^{-1}$$

$$x_5 = 7.0292257 \ 10^{-1}$$

$$x_6 = 8.7076559 \ 10^{-1}$$

$$x_7 = 9.7455395 \ 10^{-1}$$

6.5   A NONLINEAR WAVE EQUATION

The next example is the nonlinear wave equation derived from fluid flow. Consider,

$$u_t(x,t) = u \ u_x(x,t), \quad 0 < x < 1,$$
$$0 < t < T, \qquad (6-8)$$
$$u(x,0) = g(x),$$

possessing the implicit solution,

$$u(x,t) = g(x + ut).$$

The shock phenomenon always present in Eq.(6-8) is pushed far into the future by a suitable selection of $g(x)$. As the first case let $g(x) = 0.1x$. In this case the exact solution is,

$$u(x,t) = \frac{x}{t - 10}.$$

Replacing the x-derivative with a differential quadrature of order $N=7$ and integrating the resulting set of equations from $t=0$ to $t = 1.0$, we obtain the results shown in table 6.2.

Table 6.2

Approximating the Solution of

a Nonlinear Wave Equation I

t	x	computed u(x,t)	actual(x,t)
	x_1	2.570307×10^{-3}	2.570307×10^{-3}
0.1	x_4	5.050500×10^{-2}	5.050500×10^{-2}
	x_7	9.843980×10^{-2}	9.843979×10^{-2}
	x_1	2.678530×10^{-3}	2.678530×10^{-3}
0.5	x_4	5.263157×10^{-2}	5.263157×10^{-2}
	x_7	1.025847×10^{-1}	1.025846×10^{-1}
	x_1	2.827337×10^{-3}	2.827338×10^{-3}
1.0	x_4	5.555556×10^{-2}	5.555554×10^{-2}
	x_7	1.0828399×10^{-1}	1.0828376×10^{-1}

Our next, less accurate example, is to let,

$$g(x) = 0.1 \sin(\pi x),$$

having the implicit solution

$$u(x,t) = 0.1 \sin(\pi(x + ut)) \qquad (6-9)$$

which is well behaved for $0 \leq t \leq 1$.

We compute the solution of Eq.(6-9) by a Newton-Raphson technique by using as our initial approximation, the computed value obtained from the differential quadrature version of Eq.(6-8). The order of the approximation again chosen $N=7$ and the results are given in table 6.3.

Table 6.3

Approximating the Solution of

a Nonlinear Wave Equation II

t	x	computed u(x,t)	actual u(x,t)
	x_1	$8.2435190 \ 10^{-3}$	$8.2437255 \ 10^{-3}$
0.1	x_4	$9.9951267 \ 10^{-2}$	$9.9950668 \ 10^{-2}$
	x_7	$7.7428169 \ 10^{-3}$	$7.7430951 \ 10^{-3}$
	x_1	$9.4604850 \ 10^{-3}$	$9.46697773 \ 10^{-3}$
0.5	x_4	$9.8806333 \ 10^{-2}$	$9.8798155 \ 10^{-2}$
	x_7	$6.9017938 \ 10^{-3}$	$6.9041174 \ 10^{-3}$
	x_1	$1.1618795 \ 10^{-2}$	$1.1617583 \ 10^{-2}$
1.0	x_4	$9.5506268 \ 10^{-2}$	$9.5530162 \ 10^{-2}$
	x_7	$6.0877255 \ 10^{-3}$	$6.0802054 \ 10^{-3}$

6.6   SYSTEMS OF NONLINEAR PARTIAL DIFFERENTIAL EQUATIONS

Consider next the nonlinear system of equations,

$$u_t = u\,u_x + v\,u_y \quad , \quad u(x,y,0) = f(x,y),$$

$$(6\text{-}10)$$

$$v_t = u\,v_x + v\,v_y \quad , \quad v(x,y,0) = g(x,y).$$

We wish to use differential quadrature to obtain numerical values for the functions u and v. Eq.(6-10) possesses the implicit solution,

$$u(x,y,t) = f(x+ut,y+vt),$$

$$v(x,y,t) = g(x+ut,y+vt),$$

a straightforward extension of the one dimensional case. If we let

$$u_{ij}(t) = u(x_j,y_j,t),$$

$$v_{ij}(t) = v(x_i,y_j,t),$$

then by employing the differential quadrature technique to Eq.(6-10), we get the following system of coupled ordinary differential equations,

$$u_{ij}' = u_{ij}\left(\sum_{k=1}^{N} a_{ik} u_{kj}\right) + v_{ij}\left(\sum_{k=1}^{N} a_{jk} u_{ki}\right),$$

$$(6-11)$$

$$v_{ij}' = u_{ij}\left(\sum_{k=1}^{N} a_{ik} v_{kj}\right) + v_{ij}\left(\sum_{k=1}^{N} a_{jk} v_{ik}\right),$$

$$i,j = 1,2 \ldots n,$$

with the initial conditions,

$$u_{ij}(0) = f(x_i, y_j),$$

$$v_{ij}(0) = g(x_i, y_j),$$

$$i,j = 1,2 \ldots n.$$

Since the solution of Eq.(6-11) can also process a shock phenomenon for finite t, for numerical experiments we shall chose f and g to insure that the shock takes place far from the region of interest.
Let

$$f(x,y) = g(x,y) = x + y,$$

the implicit solution of Eq.(6-10) is,

$$u(x,u) = v(x,y) = \frac{(x + y)}{(1 - 2t)}.$$

Typical results are found in table 6.4.

Table 6.4

			Approximating the Solution of a Nonlinear Equation III	
t	x	y	u(x,y,y) computed	u(x,y,t) actual
	0.025	0.025	6.3615112×10^{-2}	6.365106×10^{-2}
	0.025	0.975	1.2500001	1.249999
0.1	0.500	0.500	1.2500001	1.249999
	0.975	0.025	1.2500001	1.249999
	0.975	0.975	2.4363850	2.4363848
	0.025	0.025	2.5446831×10^{-1}	2.5446023×10^{-1}
0.4	0.025	0.975	5.0001546	4.9999958
	0.500	0.500	5.0001537	4.9999958
	0.975	0.975	9.7409111	9.745531

6.7   HIGHER ORDER SYSTEMS

We have seen that a good approximation of the first derivative of a function may often be obtained in the form,

$$u_x(x_i) = \sum_{j=1}^{N} a_{ij} u(x_j), \qquad (6\text{-}12)$$

$$i = 1, \ldots, N,$$

using differential quadrature.

We now indicate how this idea can be extended to higher order derivatives.   Viewing Eq.(6-12) as a linear transformation,

$$u_x = A u,$$

we see that the second order derivative can be approximated by,

$$u_{xx} = A(Au) = A^2 u.$$

When the method is applied to this and higher derivatives, the choice of N becomes critical. The previous results and the following work show that low order approximations can be expected to yield both qualitative and quantitative information.

Let us apply the foregoing ideas to the treatment of Burger's equation.

$$u_t + u u_x = \epsilon u_{xx}, \quad \epsilon > 0,$$

with the initial condition,

$$u(0,x) = f(x).$$

Burger's equation enjoys a Riccati type property in the sense that its solutions are expressible in terms of the solution of the linear heat equation,

$$w_t = \epsilon w_{xx}, \qquad (6\text{-}13)$$

$$w(0,x) = g(x),$$

by the transformation,

$$u = -2 \epsilon w_x /w,$$

$$f(x) = 2 \epsilon g_x(x)/g(x).$$

This property will allow us to compare our numerical results with analytical solutions of Eq.(6-13). Letting

$$u_i(t) = u(x_i,t),$$

we have the set of ordinary differential equations,

$$u_i{}'(t) + u_i(t)(\sum_{j=1}^{N} a_{ij} u_j(t)) =$$

$$\epsilon \sum_{k=1}^{N} \sum_{j=1}^{N} a_{ik} a_{kj} u_j(t),$$

$$i = 1,\ldots N,$$

with the initial conditions,

$$u_i(0) = f(x_i), \quad i = 1,\ldots N.$$

6.8 LONG TERM INTEGRATION

Consider the vector equation,

$$y' = g(y), \qquad (6\text{-}14)$$

where y is an N dimensional vector. Suppose it is desired to compute $y(T)$ where T could be large.

A first order finite difference algorithm states that,

$$w(t + \Delta) = w(t) + g(w(t)) \, \Delta,$$

leading directly to a numerical scheme where we would expect $w(n\Delta) = y(n\Delta)$ $n = 0,1,\ldots,$ $\Delta > 0$. This could be a burdensome procedure leading to time consuming computation and serious stability problems which may exist in initial value problems. What is desired from the equation is a reasonable estimate of the functional values at a few grid points rather than a highly accurate determination of the entire set of values, $w(n\Delta)$.

Here we wish to examine the general application of differential quadrature as a very efficient way to compute $y(T)$. Let the points $0 = t_0 < t_1 < \ldots t_n$ be selected and the coefficient matrix a_{ij} be chosen so that,

$$y'(t_i) = \sum_{j=1}^{N} a_{ij} \, y(t_j),$$

where the determination of the coefficients has been discussed previously. Eq.(6-14) becomes,

$$\sum_{j=1}^{N} a_{ij} y(t_j) = g(y(t_i)),$$

$$i = 1,\ldots.N.$$

We can proceed in several ways. We can consider the system of equations

$$\sum_{j=1}^{N} a_{ij} y(t_j) = g(y(t_i)),$$

as a way of determining $y(t_i)$. Secondly we can use the least squares technique. Thirdly we can use a Chebyshev norm and apply linear or nonliner programming techniques.

6.9   G(Y) LINEAR

If $G(y)$ is linear, an application of least squares techniques lead to a solution of the linear system of equations. If a Chebychev norm is used linear programming techniques can be applied.

6.10   G(Y) NONLINEAR

If $G(y)$ is nonlinear, the minimization problem with a least squares procedure requires some use of successive approximation. The quasilinear technique can be very useful in this regard.

6.11   A MATHEMATICAL PROBLEM

Consider the problem,

$$x' = Ax,$$

$$x(0) = c.$$

where A is an unknown N x N matrix,c is known and the output is known at t_i. To make use of differential quadrature, we assume x'(t) is sufficiently smooth so we can write,

$$x'_i = \sum_{j=1}^{N} b_{ik} x(t_j) = A x(t_i),$$

$$i=1,\ldots N.$$

By arranging it so the times when the system is known are at the roots of the shifted Legendre polynomial PN(x) (by a suitable choice of time scales) we may determine A by minimizing,

$$\left|\left| \sum_{j=1}^{N} (b_{ij} x(t_j) - A x(t_i)) \right|\right|, \quad (6-15)$$

where $||$. $||$ denotes a suitable norm. The minimizing is carried out over all the compo- nants . If $||$. $||$ is the usual least squares criterion,the expanded form of Eq.(6-15) be- comes,

$$\min_{a_{ij}} \sum_{k=1}^{N} \left(\left(\sum_{j=1}^{N} b_{ij} x(t_j) \right)_k - \sum_{j=1}^{N} a_{kj} x(t_i) \right)^2,$$

where $(\quad)_k$ is the k th componant of $(\quad)$ and the set of x's is known.

6.12   SYSTEMS WITH PARTIAL INFORMATION

In our final discussion in this chapter we
consider the problem where only partial knowledge of the system is known ,   but the system
can be observed in a restricted sense.

Consider the partial differential equation,

$$u_t(x,t) = g(x,t,u,u_x(x,t)),   \qquad (6-16)$$

$$u(x,0) = b(x),   a \leq x \leq b, t > 0.$$

We shall assume that the initial conditions
for Eq.(6-16) are, at least, partially known
and the solution can be sampled at certain pre-
determined points within   the semi-infinite
strip, a$\leq$ x $\leq$ b, t > 0. The problem we must
solve is to determine the unknown initial conditions (and therefore the unknown solutions)
at certain points within the range of x if we
know the governing differential equation and we
can partially sample the solution.

If we again make the assumption that the
function u(x,t) satisfying Eq.(6-16) is sufficiently smooth, we can write,

$$u_x(x_i,t) = \sum_{j=1}^{N} a_{ij} u(x_j,t),   \qquad (6-17)$$

where the determination of the coefficients has
been discussed before.

Substituting Eq.(6-17) into Eq.(6-16) yields
a system of ordinary differential equations of
the form,

$$u_t(x_i,t) + g(x_i,t,u(x_i,t),\sum_{j=1}^{N} a_{ij} u(x_j,t)),$$

$$u(x_i,0) = c_j ,$$

where x_i are at the zeroes of the shifted Nth order Legendre polynomial.

Since the function g is generally nonlinear and the initial conditions may not all be known, we shall use quasilinearization techniques to establish a numerical solution to Eq.(6-17) provided u(x,t) can be partially observed or otherwise is known for some of the space points, $x_k, t_1 < T$.

By setting $u_i(t) = u(x_i,t)$, the ordinary differential equation associated with the space point x_i is,

$$u_i' = g(x_i,t,u_i,\sum_{j=1}^{N} a_{ij} u_j) .$$

For each i, we now introduce a sequence of functions,

$$u_i^{(k)}(t), \quad k = 1,2,\ldots \qquad (6-18)$$

which is generated in the following manner. If the set of functions Eq.(6-18) is known at stage k, then,

$$u_i'^{(k+1)} = g_i(u^{(k)}) + \left(\sum_{j=1}^{N} \frac{\partial\, g(u_i^{(k)})}{\partial u_j} \right)(u_j^{(k+1)} - u_j^{(k)}),$$

$$i = 1, 2, \ldots n. \qquad (6-19)$$

If an initial approximation is known, then Eq.(6-19) defines a sequence of functions for each space point i.

Because of the application of differential quadrature to the space derivative the function u represents a set of functions u_k associated with all points x_i.

As we have done before, we let,

$$u_i^{(k+1)}(t) = P_i^{(k+1)}(t) + \alpha_i h_i^{(k+1)}(t),$$

where P_i is the particular solution and h_i is the homogeneous solution associated with the point x_i. We assume that the system is such that some of the initial conditions cannot be obtained and therefore must be considered unknown .Instead of the initial conditions which are required to to obtain a solution, we assume the process $u(x,t)$ is known at M points, $x_i, i=1,\ldots M$ which can be scaled to some of the zeros of the N order differential quadrature, M<N.

The set of data functions $d_i(t)$,the observed values at x_i give partial information about the system. While the partial information will not completely describe the behavior of the system nevertheless it can be used to "fit" the mathematical model to the observed data.

We define the measurement error as,

$$\Psi = (\frac{1}{2}) \sum_{kt} \int_{1}^{T} ((\frac{\partial u}{\partial x} k) - \sum_{i=1}^{N} a_{ki} u_{i})^2 dt, \quad (6-20)$$

where u is taken over the points where the solution is unknown. Since i in Eq.(6-20) ranges over all data points, we can write,

$$\sum_{i=1}^{N} a_{ki} u_{i} = \sum_{n} a_{kn} u_{n} + \sum_{m} a_{km} (P_m + \alpha_m h_m),$$

where n ranges over all data points where the solution is known, and m ranges over the data points where the solution is unknown.

Ψ is,therefore,a quadratic form in α and can be easily minimized using the observed data to compute the required slopes as a function of time. This information can be readily found by using such techniques as the spline fit.

6.13   BIBLIOGRAPHY AND COMMENTS

Differential Quadrature is discussed in,

Bellman,R.:1970, Methods of Nonlinear Analysis, vol i & ii, Academic Press, N.Y.
see also,

Bellman,R.,B.G. Kashef and J.Casti, :1972,"Differential Quadrature: A Tecnhique for the Rapid Solution of Nonlinear Partial Differential Equations", J. Comp. Phys. 10,1, 40-52

Bellman,R. and B.G.Kashef,:1974, "Solution of the Partial Differential

Equation of the Hodgkins-Huxley   Model
using Differential Quadrature",
Math. Biosci., 19,1-8

Bellman,R.,B.G.Kashef ,R.Vasudevan and
E. S. Lee, :1975, "Differential   Quadrature and Splines", Computers and Mathematics with Applications,1,3/4,371-376

Bellman,R., G.Naadimuthu, K.M.Wang,and
E,S.Lee,:1984,"Differential Quadrature
and Partial Differential Equations"
JMAA,98,220-235

Bellman,R. and R.S.Roth,:1979,"Systems
Identification with Partial Information",
JMAA,68,2,321-333

Bellman,R. and R.S.Roth,:1979,"A Scanning
Technique for Systems Identification",
JMAA,71,2,403-411

Chapter 7
EXPONENTIAL APPROXIMATION

7.1   INTRODUCTION

A problem common to many fields is that of
determining the parameters

$$u(t) = \sum_{n=1}^{N} a_n e^{\lambda_n t}, \qquad (7-1)$$

where $u(t)$ is known.

In Chapter 5 we used the technique of differential approximation to solve this type of
problem. There, the observation was that the
right hand side of Eq.(7-1) satisfied a linear
differential equation of the form,

$$u^{(N)} + b_1 u^{(N-1)} + \ldots + b_N u = 0,$$

$$u^{(i)}(0) = c_i,$$

$$i = 1, 2, \ldots N-1.$$

Hence, in place of finding an approximate
exponential polynomial we look for an approximate
linear differential equation. We determined the
set of coefficients b_i by minimizing the functional,

$$J(u,b) = \int_0^T (u^{(N)} + b_1 u^{(N-1)} + \ldots + b_N u)^2 dt,$$

with respect to b_i. Having obtained a set of values
for the coefficients we then determined the
best c_i by another quadratic minimization.

If the values of $u(t)$ are given at discrete
time points, we used a finite difference version of the foregoing method.

In what follows we wish to present a different approximation which we may consider a natural extension of differential approximation.

7.2   APPROXIMATION IN FUNCTION SPACE

Many of the methods currently used depend upon a
suitable first approximation of a_k and λ_k.
Our suggestion is that the a_k and λ_k be replaced
by other parameters which have immediate physical
meaning. Specifically, we want to associate $u(t)$
with a particular physical process. This provides
reasonable ranges of the unknown parameters
immediately.

Since there are many possible choices, which
can be described by linear differential equations with constant coefficients, we end up
with a best fit to the process. Here, physical
intuition will very often serve as an important
guide in the selection of the process.

Each individual process yields a parameterization of a_k and λ_k.

7.3   AN EXAMPLE - PHARMACOKINETICS

Let us illustrate the foregoing with the problem of fitting a function by a set of four exponentials.   As the underlying physical process, we take a four compartmental model in pharmacokinetics. There are only two possible different structures to fit the problem.

Figure   7.1

A Compartmental Model

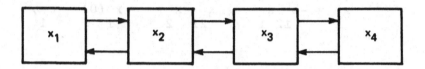

and

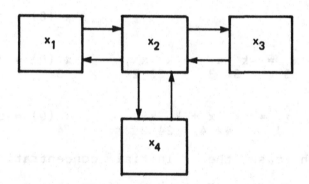

All other system are equivalent to this set.

In the first case we have the set of equations,

$$\dot{x}_1 = -k_{12} x_1 + k_{21} x_2 \quad ; \quad x_1(0) = c_1,$$

$$\dot{x}_2 = k_{12} x_1 - (k_{21} + k_{23}) x_2 + k_{32} x_3,$$

$$x_2(0) = c_2,$$

$$\dot{x}_3 = k_{32} x_2 + k_{34} x_4 - (k_{32} + k_{42}) x_2,$$

$$x_3(0) = c_3,$$

$$\dot{x}_4 = k_{34} x_3 - k_{43} x_4 \qquad\qquad x_4(0) = c_4,$$

In the second case

$$\dot{x}_1 = -k_{12} x_1 + k_{21} x_2, \qquad x_1(0) = c_1,$$

$$\dot{x}_2 = k_{12} x_1 - (k_{21} + k_{23} + k_{24}) x_2 + k_{32} x_3 + k_{42} x_4,$$

$$x_2(0) = c_2,$$

$$\dot{x}_3 = -k_{32} x_3 + k_{23} x_2, \qquad x_3(0) = c_3,$$

$$\dot{x}_4 = -k_{42} x_4 + k_{24} x_2, \qquad x_4(0) = c_4,$$

In each case the initial concentration c_i and the rate constants k_{ij}, furnish a parameterization of the exponents λ and the coefficients a_k. What is important to note is that while each choice of K_{ij} and c_i yields a set λ_k and a_k, each choice of λ_k, a_k may not correspond to a physically acceptable set K_{ij} and c_i ie. $k_{ij} > 0$, $c_i > 0$.

Thus, a procedure of successive approximations should be based on the use of differential equations with $x_1(t)$ representing u(t).

7.4   OTHER PHYSICAL PROCESSES

Let us note that the linear transport theory and electric circuit theory can also be used for the foregoing purpose. The linear differential equation would now be subject to two point boundary conditions.

7.5   PRONEY'S METHOD

Another useful method for obtaining an exponential fit to a known function is Proney's method. Using this method we seek to determine the parameters in the approximation,

$$x(t) = c_1 e^{a_1 t} + c_2 e^{a_2 t} + \ldots + c_n e^{a_n t}.$$

$$(7-2)$$

Quite a bit of information can be deduced about the structure of the process from which x(t) is derived, if Eq.(7-2) is explicitly known.

If we make the substitution,

$$P_k = e^{a_k},$$

then we can write

$$x(t) = C_1 P_1^t + C_2 P_2^t + \ldots + C_n P_n^t.$$

$$(7-3)$$

We suppose that a linear change of variable has been introduced, in advance, in such a way that the values of $x(t)$ are specified at N equally spaced points, $n = 0,1,\ldots,N-1$.

If the set of points are to fit the expression Eq.(7-3), then it must be true that,

$$c_1 + c_2 + \ldots c_N = x(0),$$

$$c_1 P_1 + c_2 P_2 + \ldots + c_N P_N = x(1),$$

$$(7-4)$$

$$c_1 P_1^2 + c_2 P_2^2 + \ldots + c_N P_N^2 = x(2),$$

$$\vdots$$

$$c_1 P_1^{N-1} + c_2 P_2^{N-1} + \ldots + c_N P_N^{N-1} = x(N-1).$$

If the constants $P_1, P_2 \ldots P_N$ are known, the set composes of N linear equations in N unknowns and can be solved exactly. If $N > n$, then the unknown constants are found approximately by least squares.

If, however, the P's also must be determined, at least 2n equations are needed which are now nonlinear in the P's. This difficulty can be minimized by the method we describe next.

Let $p_1, p_2, \ldots P_n$ be the roots of the algebraic equations,

$$P^n - a_1 p^{n-1} - a_2 p^{n-2} - \ldots - a_n = 0, \qquad (7\text{-}5)$$

where the left hand member of Eq.(7-5) is identified with the product $(p - p_1)(p - p_2)$. $\ldots (p - p_n)$. In order to determine the coefficients $a_1, \ldots a_n$, we multiply the first equation of Eq.(7-4) by a_n, the second by a_{n-1} and so on until the (n-1)st equation is multiplied by a_1.

The last equation is multiplied by -1 and the results added. If use is made of the fact that each p satisfies Eq.(7-5), the results can be seen to be of the form,

$$x(n) - a_1 x(n-1) - \ldots - a_n x(0) = 0 .$$

A set of N-n-1 equations are found in the same way starting successively with the second, third $\ldots$ n-1 equation. In this way we find that Eq.(7-4) and Eq.(7-5) imply the N-n set of linear equations,

$$x(n-1)\ a_1 + x(n-2)\ a_2 + \ldots + x(0)\ a_n = x(n),$$

$$x(n)\ a_1 + x(n-1)\ a_2 + \ldots + x(1) = x(n+1),$$

$$\vdots$$

$$x(N-2)a_1 + x(N-3)a_2 + \ldots + x(N-n-1)a_n = x(N-1).$$

Since the ordinates are known, the set of equations can be solved directly for the a's if N = 2n, or solved approximately by the method of least squares if N>2n.

After the a's are determined, the P's are found as the roots of Eq.(7-5), which may be real or complex. The equation Eq.(7-4) then becomes a set of linear equations in the coefficients c.These can be determined from any set of the equations, preferably, by applying the least squares technique to the entire set. Thus the nonlinearity of the system is concentrated in the single algebraic equation Eq.(7-5). The technique described in known as Proney's method.

7.6 THE RENEWAL EQUATION

The application of exponential approximation to the renewal equation is very instructive for it serves to transform the integral equation to a system of differential equations, Consider the renewal equation,

$$u(t) = f(t) + \int_0^t k(t - s)\ u(s)\ ds,$$

where the kernel k(t) is defined for t>0.

Using methods previously described in this chapter, we let,

$$k(t) = \sum_{n=1}^{N} a_n e^{\lambda_n t},$$

where a_n and λ_n can be readily computed.

Therefore

$$u(t) = f(t) + \sum_{n=1}^{N} a_n \int_0^t e^{\lambda_n (t - s)} u(s)ds.$$

$$(7\text{-}6)$$

We now let,

$$u_n(t) = \int_0^t e^{\lambda_n (t - s)} u(s) ds,$$

and consider the following simple analysis.

$$u_n(t) = e^{\lambda_n t} \int_0^t e^{-\lambda_n s} u(s)ds,$$

$$u_n(t) e^{-\lambda_n t} = \int_0^t e^{-\lambda_n s} u(s)ds. \qquad (7\text{-}7)$$

Differentiating both sides of Eq.(7-7) with respect to t , we have,

$$\frac{d}{dt}(u(t) e^{-\lambda_n t}) = e^{-\lambda_n t} u(t),$$

or

$$(u'_n - \lambda_n u_n - u) e^{-\lambda_n t} = 0.$$

We are led immediately to a system of linear coupled differential equations,

$$u'_n - \lambda_n u_n = u(t) \quad , \quad u_n(0) = 0,$$

and from Eq.(7-6)

$$u(t) = f(t) + \sum_{n=1}^{N} a_n u_n .$$

7.7   THE FREDHOLM INTEGRAL EQUATION

The analysis of the last section leads to a more interesting result if we consider the Fredholm integral equation.

Consider the equation,

$$u(t) = f(t) + \int_0^1 k(|t - s|)g(u(s)) \, ds,$$

where the kernel $k(t)$ is known for all $t > 0$. Now, for any value of t , we may write

$$u(t) = f(t) + \int_0^t k(t - s)g(u(s)) \, ds$$

$$+ \int_t^1 f(s-t)g(u(s))ds,$$

and since we can approximate $k(t)$

$$k(t) = \sum_{n=1}^{N} a_n e^{\lambda_n t} ,$$

we have

$$u(t) = f(t) + \sum_{n=1}^{N} a_n (\int_0^t e^{\lambda_n (t-s)} g(u(s))ds$$

$$+ \int_t^1 e^{\lambda_n (s-t)} g(u(s))ds).$$

Now we let

$$u_n(t) = \int_0^t e^{\lambda_n (t-s)} g(u(s)) \, ds$$

and

$$v_n(t) = \int_t^1 e^{\lambda_n (s-t)} g(u(s)) \, ds.$$

Following the same procedure as above, we are led to the two point boundary value problem given below.

$$u_n' - \lambda_n u_n = g(u(t)) , \quad u_n(0) = 0$$

$$v_n' + \lambda_n v_n = g(u(t)) , \quad v_n(1) = 0$$

$$u(t) = f(t) + \sum_{n=1}^{N} (u_n(t) + v_n(t)).$$

7.8   BIBLIOGRAPHY AND COMMENTS

For further study of mathematical models in pharmacokinetics, see

Bellman,R.:1970, "Topics in Pharmaco-kinetics I: Concentration Dependant Rates ",Math. Biosci.,6,1,13-17

Bellman,R.:1971, "Topics in Pharmaco-kinetics II: Identification of Time Lag Processes",Math. Biosci,11,337-342

Bellman,R.:1971 "Topics in Pharmaco-kinetics III : Repeated Dosage and Impulse Control",Math. Biosci.,12,1-5

Bellman,R.:1972, "Topics in Pharmaco-kinetics IV: Approximation in Process Space and Fitting by Sums of Exponentials", Math. Biosci.,14,3/4, 45-47

Chapter 8

THE RICCATI EQUATION

8.1   INTRODUCTION

In this chapter we wish to apply approximation techniques to the study of one of the
fundamental equations of mathematical analysis,
the first order nonlinear ordinary differential
equation.

$$u' + u^2 + p(t) u + q(t) = 0,$$

$$u(0) = c,$$

known as the Riccati equation.   This equation
plays a fundamental role in the analysis of solutions of the second order linear differential
equation,

$$w'' + a(t) w' + b(t) w = 0,$$

$$w(0) = a,$$

$$w(L) = b,$$

and thus occupies an important position in
quantum mechanics in connection with the Schro-
dinger equation. Furthermore, together with its
multidimensional analogue, it enjoys a central
place in dynamic programming and invariant em-
bedding.

8.2   THE LINEAR DIFFERENTIAL EQUATION

Consider the linear differential equation,

$$w'' + a(t) w' + b(t) w = 0. \qquad (8-1)$$

Let us introduce a change of variables,

$$w = e^{\int v \, dt}. \qquad (8-2)$$

Substituting Eq.(8-2) into Eq.(8-1), we immediately obtain,

$$v' + v^2 + a(t) v + b(t) = 0, \qquad (8-3)$$

the Riccati equation.

If the end conditions of Eq.(8-1) are given by $w(0) = a$, $w(L) = b$, then we are confronted with the solution of a two point boundary value problem which can be numerically troublesome.

Let w_1 and w_2 be the two principle solutions of Eq.(8-1) defined by the initial conditions,

$$w_1(0) = 1 \quad , \qquad w_2(0) = 0,$$

$$w_1'(0) = 0 \quad , \qquad w_2'(0) = 1,$$

then, by virtue of the linearity of Eq.(8-1), every solution can be represented as a linear combination,

$$w(t) = c_1 \, w1(t) + c_2 \, w2(t).$$

Now applying the boundary conditions,

$$a = c_1 ,$$

$$b_1 = c_1 w_1(L) + c_2 w_2(L),$$

and

$$w(t) = a_1 w_1(t) + \left(\frac{b_1 - a_1 w_1(L)}{w_2(L)} \right) w_2(t).$$

Thus if $w_2(L) \neq 0$, there is a unique solution. If $w_2(L) = 0$, there is a solution only if $b=aw_1(L)$ and in this case there is a one parameter family of solutions,

$$w(t) = \alpha w_1(t) + w_2(t),$$

where α is arbitrary. Hence existence and uniqueness are here strongly interconnected as is frequently the case.

The condition $w_2(L) \neq 0$, we recognize as the characteristic value condition associated with the Sturm-Liouville equation,

$$w'' + p(t) w' + \lambda w = 0, \qquad (8-4)$$

$$w(0) = w(L) = 0.$$

In this case we demand that $\lambda = 1$ is not a characteristic value. In other words, we insist that when $\lambda = 1$, the only solution to Eq.(8-4) is the trivial one.

Numerically this solution requires a numerical evaluation of two differential equations over the interval (0,L), storing both values continually. At the end, t = L,the two constants are evaluated and the final solution is constructed as we move from t = L, back to t = 0. (It is interesting to compare this with Potter's method in chapter (3)).

Counter to this, if we solve the Riccati equation Eq.(8-3), we must solve a first order nonlinear differential equation where the initial condition is known. The final solution,however, involves an integration as defined in Eq.(8-4). This little analysis shows that a careful study of approximation methods can involve the Riccati equation quite naturally.

8.3   DIFFERENTIAL INEQUALITIES

The concept of differential inequalities can serve us well in our in our study of approximation methods.The idea we wish to pursue, due primarily to Caplyin, is to replace a given differential equation by carefully chosen differential inequalities whose associated equations can be easily solved to yield upper and lower bounds to the solution of the original equation.

Consider the first order differential inequality,

$$\frac{du}{dt} < a(t)\,u + f(t), \qquad (8-5)$$

$$u(0) = c.$$

We can write Eq.(8-5) in the form

$$\frac{du}{dt} = a(t)\, u + f(t) - p(t), \qquad (8\text{-}6)$$

$$u(0) = c,$$

where $p(t) > 0$ for $t>0$.

Let the function $v(t)$ be the solution of

$$\frac{dv}{dt} = a(t)\, v + f(t).$$

Since we know the explicit solution of Eq.(8-6),

$$u(t) = c e^{\int_0^t a(s)ds} + e^{\int_0^t a(s)ds} \cdot \int_0^t e^{-\int_0^s a(x)dx}(f(s)-p(s))ds,$$

$$= v(t) - e^{\int_0^t a(s)ds} \cdot \int_0^t e^{\int_0^s a(x)ds} p(s)ds,$$

and since $p>0$, by assumption, $u>v$, we have successfully bounded the function $u(t)$ by $v(t)$. We will now turn our attention to the Riccati equation.

8.4   SOLUTION OF THE RICCATI EQUATION IN TERMS OF THE MAXIMUM OPERATION

Let us now show that the Riccati equation can be solved in terms of a maximum operation. We can write the Riccati equation as,

$$v' = -v^2 - p(t)v - q(t). \qquad (8-7)$$

Let us replace the quantity v^2 by its equivalent expression,

$$v^2 = \max_u (2\,u\,v - u^2),$$

then Eq.(8-7) has the form,

$$v' = -\max_u (2\,u\,v - u^2) - p(t)v - q(t),$$

$$(8-8)$$

$$= \min_u (u^2 - 2\,u\,v - p(t)v - q(t)).$$

Were it not for the minimization property, the equation would be linear.

Consider the companion equation, the linear differential equation.

$$w' = u^2 - 2\,u\,w - p(t)\,w - q(t),$$

where u(t) is now a fixed function of t. Let u and w have the same initial conditions, u(0) = w(0) = c. Let v(t), the solution of the Riccati equation, exist in the interval (0,T). A proviso concerning the existence is necessary since

the solution of a Riccati equation need not
exist at all. Consider, for example, the equation,

$$u' = 1 + u^2,$$

$$u(0) = 0,$$

with the solution, $u = \tan(t)$, for $- < t < \pi/2$.

To establish the inequality, $w > v$, we observe that Eq.(8-8) is equivalent to

$$v' < u^2 - 2uv - p(t)v - q(t),$$

for all $u(t)$, or,

$$v' = u^2 - 2uv - p(t)v - q(t) - r(t),$$

$$(8-9)$$

where $r(t) > 0$ for $t > 0$.
Let us rewrite Eq.(8-9) in the following way,

$$v' - f(t)v = g(t) , \quad v(0) = c, \qquad (8-10)$$

where

$$f(t) = -(p(t) + 2u),$$

$$g(t) = u^2 - g(t) - r(t).$$

Multiplying Eq.(8-10) by $e^{\int_0^t f(s)\,ds}$, the
integrating factor, we can solve by direct

integration,

$$\frac{d}{dt}\left(v\ e^{-\int_0^t f(s)ds}\right) = q(t)\ e^{-\int_0^t f(s)ds}.$$

Hence, the solution of Eq.(8-10) is given by the expression,

$$v(t) = c\ e^{\int_0^t f(s)ds} + e^{\int_0^t f(s)ds} \int_0^t g(s)\ e^{-\int_s^t f(r)dr}\ ds.$$

$$(8-11)$$

Since $f(s) = -(p(s) + 2 u(s))$ and $g(s) = (u-g(s) - r(s))$, substituting these into Eq.(8-11) and noting that $r(t) > 0$, $t>0$, we can readily see that,

$$v < w.$$

And therefore,

$$v(t) = \min_u \left(c\ e^{-\int_0^t (2 u(s) + p(s))ds}\right.$$

$$(8-12)$$

$$\left. + \int_0^t e^{-\int_s^t (2u(r)+p(r))dr}\ (u(s)- q(s))^2\ ds.\right.$$

Thus we have been able to obtain an analytic solution to the Riccati equation in terms of a minimum operation.

8.5 UPPER AND LOWER BOUNDS

From the representation Eq.(8-12) we can obtain upper bounds by the appropriate choice of $u(t)$. From this we can deduce the upper bounds for the solution of the second order linear differential equation,

$$w'' + p(t) w' + q(t) w = 0.$$

If $q(t) < 0$, we can obtain lower bounds by replacing v by $1/w$ and proceeding as above with the equation,

$$w' - 1 - p(t) w - q(t)w^2 = 0.$$

These techniques have been used to study the asymptotic behavior of the solution as a function of a parameter or as a function of time.

More precise estimates can be obtained if we use the general transformation,

$$v = \frac{a_1(t) + a_2(t) w}{b_1(t) + b_2(t) w},$$

since w will satisfy a Riccati equation whenever v does.

8.6   SUCCESSIVE APPROXIMATIONS VIA QUASILINEARIZATION

As we have described earlier, using the technique of quasilinearization, we are lead to the following approximation scheme for the Riccati equation.

$$v'_{n+1} = v_n^2 - 2 v_n v_{n+1} - p(t)v_{n+1} - q(t),$$

$$v_{n+1}(0) = c.$$

If $v_0(t)$ is a reasonable initial approximation, then the first approximation is found by computing,

$$v_1' = v_0^2 - 2 v_0 v_1 - p(t) v_1 - q(t),$$

$$v_1(0) = c.$$

8.7   AN ILLUSTRATIVE EXAMPLE

Consider the Riccati Equation

$$u' = - u + u^2,$$

$$u(0) = c,$$

where $|c|$ is small enough to insure u exists for all $t > 0$.

We propose to approximate the nonlinear term by the linear expression.

$$u^2 = a u.$$

The quantity a is chosen so as to minimize the expression,

$$f(a,u) = \int_0^T (u^2 - au)^2 \, dt.$$

In place of u, we use v obtained from the approximating equation,

$$v' = -v + av, \quad v(0) = c,$$
namely,

$$v = c e^{-(1-a)t}.$$

The problem is then that of minimizing the transcendental function of a,

$$f(a) = \int_0^T (c^2 e^{-2(1-a)t} - ac e^{-(1-a)t})^2 \, dt,$$

a rather complicated but manageable problem.

8.8   HIGHER ORDER APPROXIMATIONS

The problem of obtaining higher order approximations in the case of the Riccati equation is quite easy.

Consider the Riccati equation,

$$u' = -u + u^2, \quad u(0) = c,$$

where $|c| \ll 1$. A corresponding equation for u^2 is,

$$(u^2)' = 2 u u,$$

$$= 2 u (-u + u^2),$$

$$= -2 u^2 + 2u^3, \quad u(0) = c^2.$$

Writing,

$$u = u_1,$$

$$u^2 = u_2,$$

we have the system

$$u_1' = -u_1 + u_2, \qquad u_1(0) = c,$$

$$u_2' = -2u_2 + 2u_1^3, \qquad u_2(0) = c^2.$$

To obtain a linear system, we write,

$$u_1^3 = a_1 u_1 + a_2 u_2,$$

and determine the coefficients a_1 and a_2 by the condition that the expression,

$$J(u, a_1, a_2) = \int_0^T (u_1^3 - a_1 u_1 - a_2 u_2)^2 \, dt,$$

is minimized.
The calculation is carried out using the function $u_0 = v_1$, obtained as a solution of,

$$v_1' = -v_1 + v_2, \qquad v_1(0) = c,$$

$$v_2' = -2 v_2, \qquad v_2(0) = c^2.$$

Continuing in this fashion ,using higher powers of u, we obtain approximations to arbitrary accuracy.

8.9   MULTIDIMENSIONAL RICCATI EQUATION

The Riccati equation can readily be expanded to the multidimensional case by considering,

$$R' = A - R^2, \qquad R(0) = I. \qquad (8-13)$$

In this section we wish to find both upper and lower bounds on R where A is positive definite. In the more general situation,

$$R' = A + B R + R C + R D R, \qquad R(0) = C.$$

We begin the analysis of Eq.(8-13) by considering the identity,

$$R^2 = (S + R - S)^2$$

$$= S^2 + S(R - S) + (R - S)S + (R - S)^2$$

$$> S^2 + S R + R S - 2 S^2$$

$$> S R + R S - S^2,$$

for all symmetric matrices S, with equality if and only if R = S.Thus Eq.(8-13) leads to the equation,

$$R' < A + S^2 - S R - R S, \qquad R(0) = I.$$

Consider the solution of the corresponding equation,

$$Y' = A + S^2 - S Y - Y S, \quad Y(0) = I. \quad (8\text{-}14)$$

where $Y = F(S,t)$.

Now, it can be shown that the matrix equation,

$$X' = A(t) X + X B(t) + F(t), \quad X(0) = C,$$

may be written,

$$X = Y_1 C Y_2 + \int_0^t Y_1(s) Y_1^{-1}(t) F(s) Y_2^{-1}(t) Y_2(s) ds,$$

$$(8\text{-}15)$$

where
$$Y_1' = A(t) Y_1, \quad Y_1(0) = I,$$

$$Y_2' = Y_2 B(t), \quad Y_2(0) = I.$$

The implicit solution of Eq.(8-14) , as given by Eq.(8-15) allows us to conclude

$$R < F(S,t), \quad (8\text{-}16)$$

for t >0 and all S >0.

Next we make the change of variable, $R = (Z^{-1})$. The equation for Z is then

$$Z' = I - Z A Z, \quad Z(0) = I.$$

As we did above, we can conclude that,

$$Z A Z > S A Z + Z A S - S A S,$$

for all symmetric S. Hence,

$$Z' < I + S A S - (S A Z + Z A S), \quad (8\text{-}17)$$

so, $\qquad\qquad Z' < G(S,t),$

where $G(S,t)$ is the solution of,

$$W' = I + S A S - (S A W + W A S),$$

$$W(0) = I.$$

Combining Eq.(8-16) and Eq.(8-17), we have the upper and lower bounds of the matrix R, the solution to the multidimensional Riccati equation,

$$G(S_1,t)^{-1} < R < F(S_2,t),$$

for t >0, where $S_1(t)$ and $S_2(t)$ are arbitrary positive definite matrices in the interval (0,T).

8.10   VARIATIONAL PROBLEMS AND THE RICCATI EQUATION

Let us consider a multidimensional system whose behavior is governed by the variational functional,

$$J(X) = \int_a^T ((X',X')+(X,B(t) X))dt,$$

with $X(a) = C.$ Here X is an n dimensional vector and B is an n x n matrix.
Let,

$$f(C,a) = \min_{X} J(X),$$

then we obtain the equation,

$$-\frac{\partial f}{\partial a} = \min_{v} ((v,v) + (C,B(a)C) + (gradf)).$$
$$(8-18)$$

The minimizing v is given by,

$$v = -(grad\ f)/2.$$

System Eq.(8-18) reduces to,

$$-\frac{\partial f}{\partial a} = (C,B(a)C) - (gradf,gradf)/4,$$

subject to f(C,T) = 0 .
We now employ the fact that,

$$f(C,a) = (C,R(a)\ C) ,$$

as seen from the associated Euler equation. We
see that R satisfies the matrix Riccati Equation.

8.11   BIBLIOGRAPHY AND COMMENTS

Much of the results of this chapter are proven
in

Bellman,R,:1970, Methods of Nonlinear
Analysis, vol I & II ,Academic Press,
N.Y.

see also,

Bellman,R.:1978,"Quasilinearization and

the Matrix Riccati Equation',JMAA,64,1,
106-113

Bellman,R. and R. Kalaba,:1965, Quasilinearization and Nonlinear Boundary
Value Problems, American Elsevier, N.Y.

Bellman,R, and R. Vasudevan :1967,"Upper
and Lower Bounds for the Solution of the
Matrix Riccati Equation",JMAA,17,373-379

Chapter 9

SOLUTION OF APPROXIMATE EQUATIONS

9.1 INTRODUCTION

In this chapter we wish to explore some of the ideas surrounding approximate equations. The basic idea is the following: if we are confronted with a nonlinear differential equation whose solution is unknown, then we would like to replace this equation with a set, one or more, of approximating equations whose solutions are known. Our goal is to see if we can obtain an approximate solution to the nonlinear system by exact solutions to the approximating equation. Since we are, in fact, approximating one differential equation by a set of others, this chapter considers the interrelations between them.

We wish to state at the outset that nonlinear equations are notoriously difficult to solve in general and we intend to pick our way very carefully and we may skip some of the more interesting details.

9.2 FIRST ORDER DIFFERENTIAL EQUATIONS

As a preliminary exercise, the results of which we will use later, we consider the inhomogeneous differential equation,

$$u' + a(t) u = b(t), \qquad (9\text{-}1)$$

$$u(0) = c.$$

If we multiply Eq.(9-1) by the integrating factor,

$$e^{\int_0^t a(s)ds}$$

and following the procedure described in the last chapter ,we find that

$$u(t) = c\, e^{\int_0^t a(s)ds} + e^{\int_0^t a(s)ds} \int_0^t e^{\int_0^s a(r)dr} b(s)ds.$$

Thus it is quite easy to write the solution of Eq.(9-1) in terms of the function a(t) , the forcing function b(t) and the initial condition c.

9.3 THE SECOND ORDER DIFFERENTIAL EQUATION

For completeness we consider the similar analysis for the second order differential equation,

$$u'' + a(t) u' + b(t) u = g(t). \qquad (9\text{-}2)$$

Let u_1 and u_2 be two linearly independent solutions to the homogeneous equation,

$$u'' + a(t) u' + b(t) u = 0. \qquad (9\text{-}3)$$

We seek a solution to Eq.(9-2) in terms of the solutions of Eq.(9-3), and in this case we choose to use the method of separation of parameters.

We wish to determine the functions w_1 and w_2 such that,

$$u = u_1 w_1 + u_2 w_2, \qquad\qquad (9\text{-}4)$$

is a solution of Eq.(9-2).

If we differentiate Eq.(9-4),

$$u' = w_1' u_1 + w_1 u_1' + w_2' u_2 + w_2 u_2',$$

and apply the condition

$$w_1' u_1 + w_2' u_2 = 0,$$

then differentiating the results again

$$u'' = w_1 u_1'' + w_1' u_1' + w_2 u_2'' + w_2' u_2',$$

and putting the results into Eq.(9-2), we have,

$$g(t) = u'' + a(t)u' + b(t)u$$

$$= w_1 (u_1'' + a(t)u_1' + b(t)u_1)$$

$$+ w_2 (u_2'' + a(t)u_2' + b(t)u_2)$$

$$+ (w_1' u_1' + w_2' u_2').$$

Since u_1 and u_2 are the chosen solutions of Eq.(9-3) we have,

$$w_1' u_1' + w_2' u_2' = g(t),$$

$$w_1' u_1 + w_2' u_2 = 0.$$

Now we can solve for w_1' and w_2', giving

$$w_1' = \frac{-g(t)\, u_2}{(u_2' u_1 - u_1' u_2)},$$

$$\tag{9-5}$$

$$w_2' = \frac{g(t)\, u_1}{(u_2' u_1 - u_1' u_2)}.$$

We note in passing that the determinate of Eq.(9-5) is the Wronskian,

$$W(t) = \begin{vmatrix} u_1 & u_2 \\ u_1' & u_2' \end{vmatrix},$$

which is nonzero if $u_1(t)$ and $u_2(t)$ are linearly independent. Now if we consider

$$W'(t) = u_1 u_2'' + u_1' u_2' - u_1'' u_2 - u_1' u_1,$$

$$= -a(t)\,(u_1 u_2' - u_2 u_1'),$$

$$= -a(t)W(t),$$

and

$$W(t) = W(0)\, e^{-\int_0^t a(s)ds}.$$

Further if u_1 and u_2 are chosen so that

$$u_1(0) = 1, \quad u_1'(0) = 0,$$

$$u_2(0) = 0, \quad u_2'(0) = 1,$$

we can immediately evaluate w_1 and w_2 from Eq.(9-5) and find the solution of Eq.(9-2) in terms of the linearly independent solutions of Eq.(9-3). Finally,

$$u(t) = c_1 u_1 + c_2 u_2$$
$$+ \int_0^t g(s)\, e^{\int_0^s a(r)dr}\, (u_1(s)u_2(t) - u_1(t)u_2(s))ds.$$

9.4   DISCUSSION

The simple classical analysis shown above has produced analytic solutions for general first and second order linear differential equations in terms of the system parameter functions , the initial conditions and the forcing functions.

Results of this kind allows us to use general inhomogeneous equations in our quest for approximate equations for these have convenient forms of solutions. Of course, the set of equation could be expanded to any equation in which a convenient solution is known, but we will restrict ourselves to those found above.

9.5   LINEAR PERTURBATIONS

A very straight forward way to implement our idea is to consider the linear perturbation.

Let us consider first a small perturbation to the linear differential   , equation

$$u'' + (a(t)+\epsilon a_1(t)) u' + (b(t)+\epsilon b_1(t))u = 0,$$

$$(9-6)$$

$$u(0) = c_1,$$
$$u'(0) = c_2,$$

where ϵ is considered small. We expand u in a series about ϵ,

$$u(t) = u_0(t) + \epsilon u_1(t) + \epsilon^2 u_2(t) + ...,$$

$$(9-7)$$

and let the initial conditions hold uniformly in ϵ, so,

$$u_0(0) = c_1, \quad u_0'(0) = c_2,$$

$$u_i(0) = 0,$$
$$u_i'(0) = 0. \qquad i = 1,2 ...$$

Substituting Eq.(9-7) into Eq.(9-6) and equating coefficients of the powers of ϵ to zero, we obtain a series of equations.

$$u_0'' + a(t)u_0' + b(t)u_0 = 0,$$

$$u_1' + a(t)u_1' + b(t)u_1 = -(a_1(t)u_0' + b_1(t)u_0).$$

$$\vdots$$

Therefore the classical perturbation technique replaces an equation whose solution is unknown with a sequence of equations with known solutions. The value of ϵ and the behavior of the parameter functions determine how many terms must be solved to get an accurate solution.

9.6   THE VAN DER POL EQUATION I

We now consider the classical nonlinear differential equation,

$$u'' + \epsilon(u^2 - 1)u' + u = 0, \qquad (9-8)$$

where again ϵ is small. Classical analysis of this equation reveals the existence of secular, destabilizing terms, occurring in the straightforward application of the perturbation technique. This is closely associated with the fact that the nonlinearity in Eq.(9-8) not only effects the amplitude of the solution but also its frequency. To account for this, we introduce a change in the independent variable,

$$t = s(1 + c_1\epsilon + c_2\epsilon^2 + \ldots), \qquad (9-9)$$

so that
$$d/ds = (d/dt)(dt/ds)$$

$$= d/dt(1 + c_1 \epsilon + c_2 \epsilon^2 + ...).$$

We can now write;

$$\frac{d^2 u}{ds^2} + \epsilon(u^2 - 1)\frac{du}{ds}(1 + c_1 \epsilon + c_2 \epsilon^2 +..)$$

(9-10)

$$+ (1 + c_1 \epsilon + c_2 \epsilon^2 + ...)^2 u = 0.$$

Since for $\epsilon = 0$, $u = \cos s$ is a solution of
Eq.(9-9), we let

$$u = a \cos s + \epsilon u_1(s) + .. \qquad . \quad (9-11)$$

Putting Eq.(9-11) into Eq.(9-10) and setting
the coefficients of the powers of ϵ to zero,
we get for the first approximating equation,

$$\frac{d^2 u_1}{ds^2} + u_1 = -c_1 a \cos s,$$

$$(9-12)$$
$$+ (a(\frac{a^2}{4} - 1)\sin s + \frac{a^3}{4} \sin 3s).$$

We observe that the secular terms in Eq.(9-12)
are $\cos s$ and $\sin s$ both of which can be

eliminated by setting $c_1 = 0$ and $a = 2$.

Through this analysis we have achieved the following result. If we begin with the nonlinear Van der Pol equation and we ask for a periodic solution for small but finite ϵ, the approximate differential equation becomes,

$$\frac{d^2 u_1}{ds^2} + u_1 = (a^3/4)\ \sin 3s,$$

$$t = s + O(\epsilon^2), \qquad\qquad\qquad (9\text{--}13)$$

and we know the solution of Eq.(9-13) exactly.

9.7   THE VAN DER POL EQUATION II

We now wish to apply a new approximation technique to the Van der Pol equation,

$$u'' + \epsilon(u^2 - 1)u' + u = 0. \qquad (9\text{--}14)$$

Here we seek to approximate the nonlinear term in Eq.(9-14) by a linear combination of u and u', so we let,

$$(u^2 - 1)u' = a_1 u + a_2 u',$$

where a_1 and a_2 are constant.
   The measure of error in this approximation is taken to be,

$$\Psi = \int_0^{2\pi} ((u^2 - 1)u' - a_1 u^2 - a_2 u'^2) \, dt$$

and we seek a_1 and a_2 so as to minimize the error Ψ. Therefore,

$$\int_0^{2\pi} ((u^2 - 1)u' - a_1 u - a_2 u')u \, dt = 0, \quad (9\text{-}15)$$

and

$$\int_0^{2\pi} (u^2 - 1)u' - a_1 u - a_2 u')u' \, dt = 0. \quad (9\text{-}16)$$

If u is to have period 2π,

$$\int_0^{2\pi} (u^2 - 1)u \, u' \, dt = 0, \quad (9\text{-}17)$$

and

$$\int_0^{2\pi} u \, u' \, dt = 0.$$

It follows from Eq.(9-15) that $a_1 = 0$ and from Eq.(9-16),

$$a_2 = \frac{\int_0^{2\pi} (u^2 - 1)u'^2 \, dt}{\int_0^{2\pi} u'^2 \, dt} \, . \qquad (9\text{-}18)$$

Therefore the approximate equation for Eq.(9-14) is,

$$u'' + \epsilon a_2 u' + u = 0. \qquad (9\text{-}19)$$

However a_2 depends on the unknown solution, so we must further constrain the system.

Let us only consider periodic solutions so that by Eq.(9-19) $a_2 = 0$. It follows from Eq.(9-18) that,

$$\int_0^{2\pi} (u^2 - 1)u'^2 \, dt = 0,$$

and if we set $u = a \cos t$,

$$a^2 = \frac{\int_0^{2\pi} \sin^2 t \, dt}{\int_0^{2\pi} \sin^2 \cos^2 t \, dt} = 4.$$

If we had not known in advance that the period was 2π, we would have used as a measure of approximation,

$$\lim_{0} \int_{0}^{T} ((u-1)u' - a_1 u - a_2 u')^2 \, dt.$$

as $T \longrightarrow \infty$

9.8   THE RICCATI EQUATION

Consider the Riccati equation,

$$u' = -u + u^2, \qquad\qquad (9\text{-}20)$$

and apply the following linearization.
Let

$$u_k(t) = u^k, \qquad k = 1,2\ldots$$

Therefore Eq.(9-20) becomes,

$$u_k{}' = k u^{k-1} u' = k u^{k-1} (-u + u^2),$$

or

$$u_k{}' = -k u_k + k u_{k+1},$$

establishing a   direct correspondence   between
the first   order nonlinear system and   an infinite set of linear equations.

9.9   U" + A(T)U = 0

In this section, we wish to consider the second order differential equation,

$$u'' + a^2(t) \, u = 0, \qquad (9\text{-}21)$$

which arises in such areas of study as quantum mechanics, wave equations and diffusion processes.

We wish to transform Eq.(9-21) into an equation in which an approximate equation can be found . Consider the Liouville transformation. Let $a(t) > 0$ for $t > 0$ and introduce the variable,

$$s = \int_0^t a(r) \, dr, \qquad (9\text{-}22)$$

then since,

$$u' = \frac{du}{dt} = \frac{du}{ds} \frac{ds}{dt} = a(t) \frac{du}{ds},$$

Eq.(9-21) becomes,

$$\frac{d^2u}{ds^2} + \frac{a'(t)}{a^2(t)} \frac{du}{ds} + u = 0, \qquad (9\text{-}23)$$

and if we further let,

$$u = e^{-\frac{1}{2}\int_0^t p(r)\, dr}\, v,$$

this reduces the equation

$$u'' + p(t)\, u' + q(t)\, u = 0, \qquad (9-24)$$

to the form,

$$v'' + \left(q(t) - \left(\frac{1}{2}\right) p'(t) - \left(\frac{1}{4}\right) p(t)^2\right) v = 0,$$

thus eliminating the middle term in Eq.(9-24). Further, if we set

$$u = e^{-\frac{1}{2}\int_0^s \frac{a'(r)}{a(r)^2}\, dr}\, v, \qquad (9-25)$$

then noting from Eq.(9-22) that t is a function of s and applying the same transformation to Eq.(9-23) we have,

$$v'' + \left(1 - \frac{1}{2}\frac{d}{ds}\left(\frac{a'(s)}{a(s)^2}\right) + \frac{1}{4}\left(\frac{a'(s)}{a(s)^2}\right)^2\right)v = 0.$$

$$(9-26)$$

Since,

$$\int_0^s \frac{a'(r)}{a^2(r)}\, dr = \int_0^s \frac{a'(r)}{a^2(r)}\frac{ds}{dr}\, dr$$

$$= \int_0^t \frac{a'(r)}{a(r)} dr$$

$$= \log a(t),$$

the transformation Eq.(9-25) reduces to

$$u = \frac{v}{a(t)^{1/2}},$$

and we have succeeded in reducing Eq.(9-23) into an equation of the form,

$$v'' + (1 + b(s)) v = 0. \qquad (9-27)$$

The differential equation Eq.(9-27) is of particular interest for it can be further reduced to a first order Riccati equation by the transformation,

$$v = e^{-\int_0^t r(s)ds},$$

and Eq.(9-24) can be written,

$$r' = (1 + b(t)) - r^2. \qquad (9-28)$$

We now wish to modify Eq.(9-28) in such a way that it has the exact solution

$$v' = 1 + \frac{b}{2},$$

then the modified equation can be written,

$$r' = (1 + b + \frac{b^2}{4} + \frac{b'}{2}) - r^2,$$

which corresponds to the modified second order linear differential equation,

$$v'' + (1 + b + \frac{b'}{2} + \frac{b}{4})v = 0. \qquad (9\text{-}29)$$

Since we know one solution of Eq.(9-29), we can find the general solution by quadrature using Eq.(9-25).

9.10   ANOTHER APPROACH

We end this chapter by discussing yet another approach to solving approximate equations. Again we begin with a differential equation whose approximate solution we wish to find. Instead of using an approximate equation whose solution is known exactly, we look for an approximate equation whose properties reflect those of the original and seek a solution best fitting both equations in the least squared sense. We recognize that such a solution may not accurately fit the original solution over the entire range on interest and we further ask if it is possible to determine regions in which different solutions can yield more accurate solutions.

As an illustrative example, we select a nonlinear differential equation whose exact solution is known.Consider the equation,

$$u' = 1 + u^2,$$

$$u(0) = 0 \;,\; 0 < t < \pi/2, \qquad (9\text{-}30)$$

having the solution $u = \tan t$.
Let the approximating equation be,

$$u' = 1 + b(t)u,$$

$$u(0) = 0,$$

and consider the error measurement of the two systems to be,

$$\Psi = \int_0^{\pi/2} ((u' - (1+u^2))^2 + (u' - (1 + b(t) u))^2)dt.$$

$$(9-31)$$

Since we know the general behavior of Eq.(9-30), its behavior could be represented by a simple polynomial, so we let

$$u = a_1 t + a_2 t^2,$$ $$(9-32)$$

$$b(t) = b_0 + b_1 t + b_2 t^2,$$

where we have satisfied the initial condition, $u(0) = 0$. Putting Eq.(9-32) into Eq.(9-31) and minimizing over the parameters a_1, a_2, b_0, b_1 and b_2, we are lead to a nonlinear algebraic equation,

$$\frac{\partial}{\partial c_i} \Psi(c_i) = 0,$$ $$(9-33)$$

where c_i runs over the parameter set.

The set Eq.(9-33) can be solved numerically
by Newton's method and the process yields the
optimal form of the approximate equation as
well as an approximate solution. Of course, the
solution extracts the kind of information assumed in the form of Eq.(9-32) and reflects an
interesting combination of power and responsi-
bility for the analyst.

Finally we indicate what happens if we allow
the approximating differential equation to
change over subintervals in the range of interest, in our case $0 < t < \pi/2$. The functional
form of the error measurement allows us to investigate one further idea which makes this
possible. Suppose we define M subintervals $0 <$
$t < T$,at which the approximate equation is allowed to switch from one form to another. Further let us stipulate the switching can occur
at only N times in the interval, where $N < M$. We
then ask for the critical times for which the
total measurement error is a minimum.

Let

$$F_N(t_i) \quad \text{be the minimum error}$$

which has occurred in the interval $(0, t_i)$ when
N changes in the approximating equation has
been made. In each subinterval, $(t_i, t_{i+1}))$,

let,

$$\Psi(t_i) = \int_{t_i}^{t_{i+1}} \left((u'-(1+u))^2 + (u'-(1 + b_i(t)u))^2 \right) dt,$$

where b_i is the set (b_0, b_1, b_2) for interval i.
Then

$$\Psi = \sum_{i=1}^{N} \Psi(t)_i,$$

where in each interval we solve for the parameter set $(a_1, a_2, b_0, b_1, b_2)$. To insure continuity of the solution at the end points,

$$u = a_0^i + a_1^i (t-t_i) + a_2^i (t - t_i)^2,$$

where

$$a_0^i = a_0^{i-1} + a_1^{i-1} (t_i - t_{i-1}) + a_2^{i-1} (t_i - t_{i-1})^2)$$

$$a_0^0 = 0.$$

Let $E(t_i, t_j)$ be the error in approximating the nonlinear equation over the interval (t_i, t_j),

$$E(t_i, t_j) = \int_{t_i}^{t_j} ((u'-(1+u^2))^2 + (u'-(1+b(t)u)^2)^2) dt.$$

Then it must be true that,

$$F_N(t_i) = \min_{t_j < t_i} (E(t_i, t_j) + F_{N-1}(t_j))$$

$$F_N(0) = 0 \quad N = 1,2 \ldots \ldots \tag{9-34}$$

$$F_0(t_j) = \int_0^t ((u'-(1+u^2))^2 + (u'-(1+b(t)u))^2 \, dt.$$

The goal of the calculation, in the example, is to compute $F_N(\pi/2)$, using the iterative scheme Eq.(9-34). As a byproduct of the computation, the N switching times, and the approximate given by b_0, b_1, b_2, are determined in each interval.

But Eq.(9-34) is just the Fundamental Equation of Dynamic Programming discussed in chapter 2. This is the rich field of modern applied mathematics where many applications occurs and further study is recommended by the references given at the end of the chapter.

9.11   DISCUSSION

The solution of approximate equations is an interesting idea. In this philosophy we are willing to forego an accurate solution to a nonlinear differential equation and accept the exact solution of an approximate equation in its place. The conditions under which this compromise is useful are numerous . Fitting complex systems into more tractable forms are especially useful in the analysis and design of practical systems where we are more interested in gross behavior rather that detail understanding.

While the perturbation techniques discussed earlier in the chapter allow us to do this precisely,later analytical and numerical methods tend to give us more freedom. Here we are allowed to select the gross behavior we are most interested in and determine grossly their effects on the system.

Finally, we suggest that these techniques can be useful in the design of experiments where the cost of building complex apparatus is very high. Here the approximate equation plays the role of defining the structure of the experimental apparatus and the analysis yields the behavior which is to be observed.

9.12   BIBLIOGRAPHY AND COMMENTS

For perturbation techniques, we use,

> Bellman,R.:1966, Perturbation Techniques in Mathematics, Engineering and Physics,Dover, N.Y.

The ideas of approximate equations are found in

> Bellman,R.:1978, "On the Solution of Approximate Equations",Nonlinear Anal., Theory,Methods and Appl. 3,5,717-719

Numerical Techniques are discussed in,

> Bellman,R. and Roth,R.:1969, "Curve Fitting by Segmented Straight Lines", J. Amer. Stat. Assoc. 64, 1077-1084

> Roth,R.:1966, "Data Unscrambling : Studies in Segmental Differential Approximation", JMAA,14, 5-22

For an introduction to Dynamic Programming, see

> Bellman,R.:1957, Dynamic Programming, Princeton University Press, Princeton, N.J.

> Bellman,R. and Dryfus,S.:1963, Applied Dynamic Programming,

Princeton University Press, Princeton, N.J.

Chapter 10

MAGNETIC FIELD DETERMINATION

10.1 INTRODUCTION

We have waited until the last chapter to
discuss ,in detail, a problem using the finite
element method as an approximation technique.
The problem with which we will be concerned is
the determination of the magnetic field arising
from a set of permanent magnets arbitrarily
placed in space containing material of varying
permeability.

We have chosen to solve this problem using
finite elements following the work of Zienkiewicz, with one exception. We have indroduced a
27 node brick formulation instead of the standard 20 node model.This means that within each
solid element a quadratic variation is allowed
which has been described in chapter 2 and is
motivated by the need to have a complete set of
polynomials with which to work. Chapter 1 discusses the necessity of having a complete set
as a prerequisite of a good approximation.

Finally the finite element method can be
used also to describe the magnets themselves,
making the method not only suitable as a technique for solving difficult three dimensional
problems successfully, but also as a way of defining the forcing environment.

10.2   THE THEORETICAL PROBLEM

A current moving in space will induce a magnetic field in the surrounding media.   The Biot-
Savart law characterizes this fact by stating
that the magnetic field, in the neighborhood of
a long straight conductor carrying a steady
current density, J , is given,

$$dB = \frac{\mu}{4\pi} \; J \; dl \; sin\theta/ \; r^2,$$

where dB is the increment of the magnetic field
located r, θ away from a conductor of length dl
having a steady current density J and a permeability μ in the surrounding media.

Figure 10.1

Incremental Biot-Savart Law

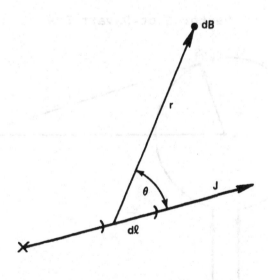

In a three dimensional space, the current
density J is considered to be a vector whose
magnitude is J and whose direction is in the
direction of the conductor, ie. in the direction of the wire in a wire wound conductor and
is directed in the direction of positive current flow. In vector notation, the Biot-Sa-
vart law can be written,

$$B(r) = \frac{\mu}{4\pi} \int_{v'} \frac{J \times (r - r')}{|r - r'|^3} \, dv,$$

where μ is constant, r is the position vector
to the point where B is to be evaluated and r'
is the position vector to dv'.

Figure 10.2

Vector Biot-Savart Law

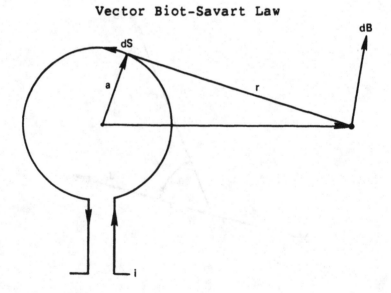

In electromagnetic theory, the vector field B is defined in terms of another field H such that,

$$B = \mu H,$$

where H is the magnetic field strength.

So for constant μ, we can write,

$$H(r) = \frac{1}{4\pi} \int_{v'} \frac{J \times (r - r')}{|r - r'|^3} dv'.$$

The problem is then clear. If we are given a conductor of volume v' located in a fixed coordinate system in space and a current density vector J known throughout v', and if we insert a known material of permeability μ in the space, then we must determine the resulting magnetic field in required areas of the space.

10.3 MAXWELL'S EQUATIONS

The basic electro-magnetic phenomenon is described by Maxwell's equations , which in vector form are,

$$\nabla \times H = \partial D/\partial t + J,$$

$$\nabla \times E = - \partial B/\partial t,$$

$$\nabla \bullet B = 0,$$

$$\nabla \bullet D = \rho,$$

$$B = \mu H, \quad D = \epsilon E.$$

where a consistent set of units is used.

In this particular class of problems, we shall consider the case of steady state conditions ($\partial/\partial t$ = 0) and no charge density (ρ =0). The permeability is assumed(μ = 1) in air and at worst μ = μ (x,y,z) in the material. (It is always assumed known.) Then the magnetic equations, equations are,

$$\nabla \times H = J, \qquad (10\text{-}1)$$

$$\nabla \cdot B = 0, \qquad (10\text{-}2)$$

$$B = \mu H. \qquad (10\text{-}3)$$

Under the assumptions above,the magnetic equations (10-1 - 10-3) are a set of linear partial differential equations. Taking advantage of this fact, we let,

$$H = H_p + H_m,$$

and using Eq.(10-1) we set,

$$\nabla \times H_p = J, \qquad (10\text{-}4)$$

$$\nabla \times H_m = 0, \qquad (10\text{-}5)$$

where H_p is the magnetic field set up by the presence of a current density J and H_m represents the perturbed field due to the presence of the permeable material. Without specifically showing it, the solution of of Eq.(10-4) is,

$$H_p(r) = \frac{1}{4\pi} \int_{v'} \frac{J \times (r - r')}{|r - r'|^3} dv',$$

which may be computed for any point r in space
since J is known throughout the volume v' of
the conductor.

Eq.(10-2) is more interesting . Since we
know from vector algebra that,

$$\nabla \times \nabla\phi = 0,$$

for any continuous scalar function, ϕ, we chose
to let

$$H_m = \nabla \phi,$$

meaning Eq.(10-5) is automatically satisified.
Hence are left with Eq.(10-2) to contend with,
or

$$\nabla \bullet \mu \nabla\phi + \nabla \bullet \mu H_p = 0, \qquad (10-6)$$

which is a partial differential equation in the
unknown scalar function ϕ .

If μ is a constant, since the vector operator is,

$$\nabla = \partial/\partial x\ i + \partial/\partial y\ j + \partial/\partial z\ k,$$

where vectors i,j,k form a basic ortho-normal
set of coordinate vectors, then Eq.(10-6) becomes,

$$\nabla^2 \phi = -(\ \partial(H_{p_x})/\partial x + \partial(H_{p_y})/\partial y + \partial(H_{p_z})/\partial z),$$

which we all know as the Poisson equation.

10.4   A VARIATIONAL PRINCIPLE

The finite element method depends on a variational principle which must be consistent with
the underlying differential equation if the
method is to be used in seeking a solution. Using this approach, we now reconsider the problem by stepping back and reformulating the
problem from a broader point of view. We seek
to find a functional involving the whole problem whose minimum, in some sense, will recap-
ture in basic differential equation, in this
case Eq.(10-6).

Consider, for the moment, the expression,

$$X = \int_v \int_0^{|H|} \mu(|H|)|H| \, d|H| \, dv - \int_s Bn \, \phi ds,$$

where v is the volume of the whole problem, s is
the bounding surface and Bn is the prescribed
value of the normal componant of the flux density on the boundary. Here we allow the perme-
ability $\mu = \mu(|H|)$ where $|H|$, defined below,
the square of the magnetic field strength.

The functional X is now minimized in the
sense that its first variation must vanish, ie.
$\delta\phi = 0$. Hence,

$$\delta X = \int_v (\mu (|H|)|H|)d\delta|H| \, dv - \int_s Bn\delta\phi ds.$$

Now, we note that $\delta H_p = 0$, and since,

$$(|H|)^2 = (H_x^2 + H_y^2 + H_z^2),$$

$$|H|\delta|H| = H_x \; \delta\partial\phi/\partial x + H_y \; \delta\partial\phi/\partial y$$

$$+ H_z \; \delta\partial\phi/\partial z,$$

$$= H \bullet \nabla\delta\phi .$$

Hence,

$$\delta X = \int_v \mu H \bullet \nabla\delta\phi \; dv - \int_s B_n \delta\phi \; ds,$$

but by Green's theorem,

$$\int_v \mu H \bullet \nabla\delta\phi dv = \int_s \mu H_n \delta\phi \; ds$$

$$-\int_v \nabla\mu H \; \delta\phi \; dv.$$

Combining results, we get,

$$\delta X = - \int_v \nabla\mu H \delta\phi \; dv + \int_s (\mu H_n - B_n)\delta\phi ds = 0.$$

$$(10-7)$$

Since Eq.(10-7) must hold for any $\delta\phi$ defined in v and on s, then,

$$\nabla\mu H = \nabla\mu(H_p + \nabla\phi) = 0,$$

$$(10-8)$$

$$\mu(H_p + \partial\phi/\partial n) = B_n \quad \text{on s.} \qquad (10-9)$$

But Eq.(10-8) is precisely the basic equation (10-6) and Eq.(10-9) tells us something about the boundary conditions. In magnetostatic problems generally all quantities tend to zero at infinity. However Eq.(10-9) defines allowable natural boundary conditions or force boundary conditions of the form ϕ given on the boundary.

10.5   THE FINITE ELEMENT METHOD

Now let us consider Eq.(10-7) more carefully,

$$-\int_v \nabla\mu H \ \delta\phi \ dv + \int_s (\mu H_n - B_n)\delta\phi \ ds = 0,$$

or

$$-\int_v \nabla(\mu(H_p +\nabla \phi))\delta\phi dv + \int_s (\mu(H_p +\partial\phi/\partial n)-B_n)\delta\phi ds.$$
$$= 0. \qquad\qquad (10-10)$$

More precisely, we can say that ,among all functions ϕ ,the one minimizing Eq.(10-10) for all $\delta \phi$, will satisfy Eq.(10-6) everywhere in v.

The finite element method is a theory of approximation in a dual sense of the word. First we break the problem into smaller physical elements and then we look for an approximate form of the solution in each element. Since the basic problem Eq.(10-6) is linear, we can do this. Within the physical element, we select a set of functions, called element shape functions, within whose bounds we agree to work. If we are lucky enough to include the true solution within the set, then Eq.(10-10) will find it. If we're not so lucky, Eq.(10-10) will the best approximation.

Now we select a set of shape functions $h_i (\xi,\eta,\zeta)$ defined over a unit cube. We choose 27 such functions which are orthogonal over the node points of the unit cube (see Eq.(2-9)).

$$h_i (\xi,\eta,\zeta) = \delta_{ij} ,$$

The physical element is the defined by the transformation,

$$x = \Sigma\, x_i\, h_i(\xi,\eta,\zeta),$$

$$y = \Sigma\, y_i\, h_i(\xi,\eta,\zeta), \quad i = 1,2,\ldots,27,$$

$$(10\text{-}11)$$

$$z = \Sigma\, z_i\, h_i(\xi,\eta,\zeta),$$

where x_i, y_i, z_i locate the 27 node points of the chosen finite element. We chose to perform the analysis by using an isoparametric representation of the unknown scalar function, ϕ. Let,

$$\phi = \Sigma\phi_i\, h_i(\xi,\eta,\zeta), \quad (10\text{-}12)$$

with the corresponding variations

$$\delta\phi = \Sigma\, \delta\phi_i\, h_i(\xi,\eta,\zeta),$$

where ϕ_i are the solution constants of the problem and $\delta\phi_i$ are arbitrary.

It follows from Eq.(10-12) that in vector form,

$$\nabla\phi = (\partial/\partial x(\phi_1 h_1 + \phi_2 h_2 + \ldots + \phi_{27} h_{27})\, i$$

$$+ (\partial/\partial y(\phi_1 h_1 + \phi_2 h_2 + \ldots + \phi_{27} h_{27})\, j$$

$$+ (\partial/\partial z(\phi_1 h_1 + \phi_2 h_2 + \ldots + \phi_{27} h_{27})\, k,$$

and therefore

$$\nabla\phi \;=\; \sum_i \phi_i \nabla h_i .$$

Now referring back to Eq.(10-10),we have

$$\int_v \mu_p (H + \nabla\phi) \bullet \nabla\delta\phi \; dv - \int_{sn} B\delta\phi ds = 0,$$

or

$$\int_v \mu_p (H + \sum_i \phi_i \nabla h_i) \bullet \sum_j \delta\phi_j \nabla h_j \; dv$$

$$\hspace{6cm} (10\text{-}13)$$

$$-\int_s B_n \sum_i \delta\phi_i \nabla h_i \; ds = 0.$$

Let us consider a solution for which ϕ and $\delta\phi$ are zero on the boundary of the problem, that is ϕ_i and $\delta\phi_i$ vanish at the boundary nodes.   Then Eq.(10-13) reduces to,

$$\int_v \mu_p (H + \sum_i \phi_i \nabla h_i) \bullet \sum_j \delta\phi_j \nabla h_j \; dv = 0,$$

or

$$\sum_j \delta\phi_j (\sum_i \phi_i \int_v \nabla h_j \bullet \mu \nabla h_i \; dv$$

$$+ \int_v \mu_p H \bullet \nabla h_j \; dv) = 0,$$

which must hold for all $\delta\phi_j$ at the interior points of the magnetic volume under study. Therefore,

$$\sum_i \phi_i \int_v \mu \nabla h_i \bullet \nabla h_j \; dv + \int_v \mu_p H \bullet \nabla h_j \; dv = 0.$$

But this is just a linear set of algebraic equations of the form,

$$D\phi + f = 0,$$

where

$$\phi = (\phi_1, \phi_2, \ldots \phi_{27}),$$

$$d_{ij} = \int_v \mu \nabla h_i \bullet \nabla h_j \, dv,$$

$$f_j = \int_v \mu H_p \bullet \nabla h_j \, ds.$$

Several points now present problems. The scalar shape functions are defined over the unit cube while the volume in Eq.(10-13) is over space. But we know the governing transformation from the cube to the element Eq.(10-11). Hence, if the functions and their first derivatives are continuous, then,

$$dv = |J| \, dv',$$

where

$$J = \begin{vmatrix} \partial x/\partial \xi & \partial x/\partial \eta & \partial x/\partial \zeta \\ \partial y/\partial \xi & \partial y/\partial \eta & \partial y/\partial \zeta \\ \partial z/\partial \xi & \partial z/\partial \eta & \partial z/\partial \zeta \end{vmatrix},$$

and $|J| \neq 0$ at all points in the unit cube . With knowledge of H_p through the unit cube, ie. throughout the element we may compute the matrix D and the vector f for any element in the problem.

Finally how do we assemble the element matrices to solve the complete problem? If we look at matrix D, we can interpret it to be an

influence matrix, that is, the matrix tells us
how the value of φ at node i is related to the
other nodes in the adjacent elements. Since the
problem is linear,if we carefully   write out
all the influences of all the nodes on nodes
for the entire problem,we can recombine the results in one gigantic matrix of the form,

$$A\phi + F = 0,$$

where A is a known system matrix, φ is the
vector of nodal values of φ and F is
a vector of known values of the forcing
function due to H_p ,

So we can now form the solution by noting,

$$\phi = A^{-1} F,$$

$$\phi = \Sigma_i \phi_i h_i ,$$

$$Hm = \nabla\phi,$$

$$= \partial\phi/\partial x\ i + \partial\phi/\partial y\ j + \partial\phi/\partial z\ k,$$

$$H = H_p + H_m .$$

10.6   COMPUTATIONAL   ASPECTS

Computationally the procedure   follows very
closely the newly developed 27 node brick used
in the finite element proceedure for structural
analysis. The shape functions are identical as
are the methods of differentiation and numerical integration. The assembly of the influence
matrix as well as its solution follow precisely
the same pattern as before .

The differences involve a new procedure for computing the elements of the influence matrix and the computation of the particular solution. A finite element model of the conductor together with the distribution of the steady state current vector J is required. The permeable material must also be modelled in a similar manner together with the permeability coefficient which must be defined throughout the material. Finally the surrounding air must be included in the finite element formulation.

10.7 ANALYTICAL CONSIDERATIONS

Based on the Maxwell's Field Equations for the electro-magnetic phenomenon, the magnetic potential ϕ is governed by the the partial differential equation

$$\nabla^2 \phi = - \nabla \bullet H_p,$$

where the forcing function H is given by,

$$H_p = \frac{1}{4\pi} \int \frac{J \times (r - r')}{|r - r'|^3} dv',$$

where J is the current vector density in the conductor v', r is the position vector of the point where H is to be formed and r' is the position vector to dv'.

The finite element approximation is in the form of a matrix equation,

$$K\phi + g = 0, \qquad\qquad (10\text{-}14)$$

where

$$K_{ij} = \int_v \mu \nabla f_i \bullet \nabla f_j \, dv,$$

$$g_i = \int_v \mu \nabla f_i \bullet H dv.$$

Here v is the volume of the problem and f_i is the set of 27 orthogonal functions for the new 27 node finite element brick formulation. We note that in equation Eq.(10-14),

$$\nabla f_i \bullet \nabla f_j = \frac{\partial f_i}{\partial x} \frac{\partial f_j}{\partial x} + \frac{\partial f_i}{\partial y} \frac{\partial f_j}{\partial y}$$

$$+ \frac{\partial f_i}{\partial z} \frac{\partial f_j}{\partial z} ,$$

$$\nabla f_i \bullet H = \frac{\partial f_i}{\partial x} H_x + \frac{\partial f_i}{\partial y} H_y + \frac{\partial f_i}{\partial z} H_z,$$

where integration is over the volume v of the conducting media.

The finite element method refers all integrations to an unit cube and a correspondence is set up such that,

$$x = \sum_i x_i f_i (\xi, \eta, \zeta),$$

$$y = \sum_i y_i f_i (\xi, \eta, \zeta),$$

$$z = \sum_i z_i f_i(\xi,\eta,\zeta),$$

where (x_i, y_i, z_i) are nodal coordinates and f_i is defined in the unit cube. The volume increment is,

$$dv = |J| \, d\xi d\eta d\zeta,$$

and

$$\left| \begin{matrix} \partial f_i/\partial x \\[2mm] \partial f_i/\partial y \\[2mm] \partial f_i/\partial z \end{matrix} \right| = (J^T)^{-1} \left| \begin{matrix} \partial f_i/\partial \xi \\[2mm] \partial f_i/\partial \eta \\[2mm] \partial f_i/\partial \zeta \end{matrix} \right|$$

where J is the Jacobean matrix.

The forcing vector H_p is referred to the unit cube in the following way,

$$H_p = H_x i + H_y j + H_z k,$$

and

$$H_x = \sum_i a_i f_i(\xi,\eta,\zeta),$$

$$H_y = \sum_i b_i f_i(\xi,\eta,\zeta),$$

$$H_z = \sum_i c_i f_i(\xi,\eta,\zeta),$$

where a_i, b_i, c_i are the nodal values of the componants of the forcing vector. The finite element formulation of the problem becomes,

$$K \phi = -g,$$

where

$$K_{ij} = \int \left(\frac{\partial f_i}{\partial x} \frac{\partial f_j}{\partial x} + \frac{\partial f_i}{\partial y} \frac{\partial f_j}{\partial y} + \frac{\partial f_i}{\partial z} \frac{\partial f_j}{\partial z} \right) |J| d\xi d\eta d\zeta,$$

$$g = \int \left(\frac{\partial f_i}{\partial x} \Sigma a_i f_i + \frac{\partial f_i}{\partial y} \Sigma b_i f_i + \frac{\partial f_i}{\partial z} \Sigma c_i f_i \right) |J| d\xi d\eta d\zeta.$$

The function ϕ is the vector of nodal values of the magnetic potential.

10.8   BOUNDARY CONDITIONS

The mathematical problem with which we are concerned is finding the solution of the Poisson equation,

$$\nabla^2 \phi = - \nabla \bullet Hp,$$

within a volume V bounded by a surface S. Theoretically this partial differential equation requires boundary conditions of the form,

$$\phi$$

specified on S.

$$\partial \phi / \partial n$$

We now ask how these boundary conditions are reflected in the finite element method. Often boundary conditions are glossed over in the eagerness to solve the equation, yet it is perhaps possible to reduce computer time by cleverly choosing nonstandard boundary conditions. We recall that ϕ is a scalar potential whose gradient represents a perturbed magnetic field due to the presence of magnetic material in space. Far away from the material, the field should vanish and it is reasonable to assume ϕ = 0 on the far boundaries. Yet this could require a great number of elements to properly model the problem. If the particular problem contains surfaces across which the solution is known to be symmetric, the size of the problem can be cut drastically by replacing part of the model by the condition, $\partial\phi/\partial n$ = 0 , on these symmetric surfaces.

The variational principle on which the finite element is based is the following,

$$\Pi = \int_v (\int_0^{|H|} \mu(|H|)d|H|dv - \int_{Sb} \overline{B}_n \phi \, ds - \int_{S\phi} B \overline{\phi}_n \, ds ,$$

$$(10\text{-}15)$$

where

$$H = H_{cp} + \nabla \phi,$$

$$B = \mu (H_p + \nabla \phi),$$

$$|H|^2 = H_x^2 + H_y^2 + H_z^2 ,$$

Sb = surface on which B_n is given,

$S\phi$ = surface on which ϕ is given.

Eq.(10-15) represents the magnetic energy stored in the field, minus that flowing through the boundaries. Basically we claim that among all φ satisfying φ = φ given on Sφ, the one for which the first variation of Π vanishes, satisfies Eq.(10-6). Taking the variation of Eq.(10-15),we have

$$\delta\Pi = \int_v (\mu|H|d|H|dv - \int_{Sb} \overline{B}_n \delta\phi ds.$$

Since φ is known on Sφ, its variation over that surface is zero and contributes nothing to the total variation. From the definition of |H|, we see,

$$|H|\delta|H| = H_x \delta H_x + H_y \delta H_y + H_z \delta H_z,$$

but by definition, $H = H_p + \nabla\phi$ and $\delta H = \nabla\delta\phi$. Therefore,

$$\delta\Pi = \int_v (\mu H \bullet \nabla\delta\phi \, dv - \int_{Sb} \overline{B}_n \delta\phi ds = 0.$$

$$(10-16)$$

Let us say a word about Eq.(10-16). The field H is assumed to be the true but unknown vector field in v, δφ is an arbitrary continuous function defined on V and zero on S φ. The true vector field, H = Hp + ∇ φ ,where Hp is known everywhere and the scalar potential is known only on S φ. In the problems for which we wish to apply the finite element method, the space of interest is embedded with finite volumes of mgnetic material. Hence μ the permuability is discontinuous on certain surfaces throughout v.

Equation Eq.(10-16) states that if H is the true field, then for any function δφ satisfying δφ = 0 on S φ, δΠ = 0. This allows us to construct the finite element method. However this

assumes Eq.(10-16) is consistent with the
governing differential equation Eq.(10-6).
Classically this consistency can be shown thru
Green's theorem which states that if a vector
function is continuous throughout a volume v ,
then,

$$\int_v \nabla \bullet f \, dv = \int_S f \bullet ds.$$
$$_n$$

Assuming continuity and integrating
Eq.(10-16) by parts, applying Green's theorem
shows Eq.(10-16) to be,

$$\int_v \nabla \bullet (\mu(H_p + \nabla\phi)) \delta\phi \, dv - \int_{Sb} (B_n - \mu H_n) \delta\phi ds,$$

$$(10-17)$$

but Eq.(10-17) must hold true for all $\delta\phi$,
therefore,

$$\nabla\mu\nabla\phi = - \nabla \bullet H_p ,$$

$$\phi = \bar{\phi} \text{ on } S\phi$$

$$\mu H_n = \bar{B}_n \text{ on } Sb .$$

But what can we say about those irritating
discontinuities caused by sudden changes in the
permuability? First, let us agree to model the
problem so that the surfaces of discontinuity
lie on the faces of the finite elements. It
follows directly from Maxwell's equations that
the normal componant of B is continuous across
the interface while the tangential componant of
H is continuous there. Within the context of
the finite element solution,it is an open question as to how to handle these discontinuities.
Since the solution is obtained at a finite number of points on such a surface, the effect is

simple neglected. However, if the flux leakage
over a material-air interface has to be computed, this question will arise again.

10.9   DISCUSSION

The finite element approximation affords us
a neat way to solve complex problems in partial
differential equations.   These techniques are
being successfully applied to a large class of
problems including,   among others,   elasticity,plasticity,heat transfer and fluid mechanics.

The finite element technique is truely a
modern method of approximation,   having roots
squarely in the   Galerkin procedure.   However,like all approximations methods, care must
be used in applying them. Defining the finite
element model to include all the needed flex-
ibility is by no means straightforward. Yet
careful construction can lead to impressive results leading us to conclude that the final
product is well worth the effort.

10.10   BIBLIOGRAPHY AND COMMENTS

There is a voluminous amount of work done on
the finite element method, we cite only,

          Gallagher,R. H. :1975, Finite
          Element Analysis Fundamentals,
          Prentice-Hall,Inc. Englewood
          Cliffs, N.J.

          Zienkiewicz,O.C.,:1971,The Finite
          Element Method in Engineering
          Science,McGraw-Hill, London

The basic ideas of this chapter are found in,

Zienkiewicz,O.C.,J.Lyness and D.R.J. Owen,:1977,   "Three Dimensional Magnetic Field Determination using a Scalar Potential- A Finite Element Solution", Trans. on Magnetics, Mag-13,5,1649-1656

The physics of magnetic   fields are well expressed in,

Halliday D. and R. Resnick,:1966, Physics parts I and II John Wiley & Sons Inc, N.Y.

The magnetic problem can also be formulated in terms of a vector potential. For details,   we suggest,

Csendes,Z.J., J. Weiss and S. R. H. Hoole,:1982, "Alternative Formul- ation of 3-D Magnetostatic Field Problems", IEEE Trans. on Magnetics, Mag-18,2,367-372

Kotiuga,P.R., and P. P. Silvester,1982, "Vector Potential Formulation for Three Dimensional Magnetostatics" , J. Appl. Phy. 53, 11, pt 2, 8399 - 8401

Mohammed,O.A.,W.A.Davis, B. D. Popovic, T. W. Nehl and N. A. Gemerdash, :1962, "On the Uniqueness of Solutions of Magnetostatic Vector-Potential Problem by Three Dimensional Finite Element Methods",J. Appl. Phy. 53,11, 8402-8404

INDEX